Problems of Conception

PROBLEMS OF CONCEPTION

*Issues of Law, Biotechnology,
Individuals and Kinship*

Marit Melhuus

Berghahn Books
New York • Oxford

Published in 2012 by

Berghahn Books

www.berghahnbooks.com

© 2012 Marit Melhuus

All rights reserved. Except for the quotation of short passages for the purposes of criticism and review, no part of this book may be reproduced in any form or by any means, electronic or mechanical, including photocopying, recording, or any information storage and retrieval system now known or to be invented, without written permission of Berghahn Books.

Library of Congress Cataloging-in-Publication Data

Melhuus, Marit.

Problems of conception : issues of law, biotechnology, individuals and kinship / Marit Melhuus. -- 1st ed.
   p. cm.
   Includes bibliographical references and index.
   ISBN 978-0-85745-502-4 (hardback : alk. paper) -- ISBN 978-0-85745-503-1 (ebook : alk. paper) 1. Conception. 2. Reproduction--Social aspects--Norway. I. Title.
QP251.M435 2012
573.609481--dc23

2011052020

British Library Cataloguing in Publication Data

A catalogue record for this book is available from the British Library.

Printed in the United States on acid-free paper

ISBN 978-0-85745-502-4 (hardback)

ISBN 978-0-85745-503-1 (ebook)

Front cover:
Edvard Munch: *Madonna* (1895-1902), details
© The Munch Museum / The Munch-Ellingsen Group / BONO, Oslo 2011
Photo: © The Munch Museum

*In memory of my mother,
Else Melhuus (1917–2010)*

Contents

Preface and Acknowledgements ix

1. Framing the Issues 1
2. Children of One's Own 23
3. Better Safe than Sorry: Legislating Assisted Conception 47
4. The Inviolability of Motherhood 71
5. The Sorting Society: Knowledge, Selection, Ethics 89
6. Concluding Reflections: Legal (Un)Certainties 109

Postscript: Some Notes on Methodology 121

Appendix: Fertility Rates, Trends and Policies in Norway 131

Notes 135

References 153

Index 167

Preface and Acknowledgements

This book has been in the making for quite a while: not only has the research underpinning it been ongoing for many years but it also draws on papers – some published, some not – that I have had the opportunity to present and discuss in many different settings. My ideas and perspectives have evolved in active dialogue with many others, and in tandem with the development and incorporation of reproductive technologies in Norway. With regard to the latter, this has been especially challenging, as my field was moving along and shifting ground while I was in midstream, so to speak. The research itself was in different ways caught up in the processes encompassing the very object of study. This has been advantageous – as well as frustrating. On the one hand it has implied a redirection of my anthropological gaze; it has also enhanced a reflexive mode, including an increased awareness of the expansion of my field – and as a consequence, the need to impose some limitations. This task was, on the other hand, not an easy one, and although a cutting off point is never arbitrary, it does entail a circumscription that obviously can be questioned. When you bring something in, something else is inevitably left out.

When I started my research on kinship and assisted conception, the public debates turned on the question of donor anonymity with regard to sperm donation (should the anonymity clause be rescinded?) and whether it should be legal to combine donor sperm with in vitro fertilization. The former had, primarily, to do with the significance of knowledge of biological origin; the latter questioned whether such a combination was just too much meddling with nature (*tukling med naturen*). As I was finalizing this book, the Norwegian *offentlighet* (that is, the public domain) was concerned – and in part outraged – by the practice of surrogacy and not least the use of Indian surrogate mothers. Surrogacy was barely on the agenda ten years ago; it was not yet conceived as a problem or a practice that Norwegians would be prone to make use of. Today, according to news reports, every week a child is born to a Norwegian – male, female or couple – through an Indian surrogate. Various surrogacy arrangements, made possible by an expanding global fertility market, are now part of Norwegian procreative practices. This is indeed a significant change, and the responses to these developments have been many, some of which I explore in this book.

There is no doubt that the timing of my research has been exceptionally rewarding and conducive to the issues that I wanted to explore. I did not know (and of course could not know) that I was tapping into a field that was developing at a pace that implied a heightened public concern. Nor could I have imagined the intensity of media coverage at certain junctures. Certain kinds of reproductive events, it seems, lend themselves to dramatic effects, intensifying the scope and magnitude of reactions. Such media coverage provided me with a broad range of positions and perspectives that focused a specific Norwegian take on biopolitical and bioethical issues, not necessarily directly related to kinship and assisted conception. One of these is the notion of a 'sorting society' that surfaced in conjunction with discussions of pre-natal diagnosis. It seemed to me that this term worked in interesting ways and that it contained a potential to yield significant anthropological insights. However, exploring this potential took me down paths that I had not initially envisioned.

The research upon which this book is based started in 1998, with a project conceived jointly with Signe Howell and Olaf Smedal entitled 'Kinship – Quo Vadis? Meanings of Kinship and Procreation in Norway'.[1] Recognizing the significance of kinship in Norwegian society and the paucity of such studies in Norway, our aim was double: to contribute to an understanding of Norwegian society, and to contribute to contemporary kinship studies. Despite the fact that there is a long tradition among Norwegian anthropologists of doing fieldwork 'at home', kinship as a focus was conspicuous by its absence. Hence, we hoped to fill a gap, as it were, by bringing kinship 'home'. Prompted by the so-called new kinship studies, and with a comparative aim, our focus was on two contemporary phenomena: assisted conception through the use of reproductive technologies, and transnational adoption. While my interest was in the former, Signe Howell pursued the latter. These combined efforts proved particularly productive as we probed and configured this universe of 'unnatural procreation'. So much so, I believe, that the results of both projects would have been otherwise if not for these joint and parallel investigations. As our work proceeded it became apparent that adoption and assisted conception are mutually constituted, and one could not be sufficiently grasped without reference to the other. My thanks to Signe Howell for many inspiring conversations and mutual collaboration throughout these years.

In 1999 I was elected Dean of the Faculty of Social Sciences at the University of Oslo, a position I held for a period of three and half years. This was not part of any plan, and inevitably my duties as Dean impinged on my research. Nevertheless, and perhaps fortuitously, much of my fieldwork could be carried out in the evenings and at weekends, as many of those I wanted to talk to also worked and were not available during the day. At this time our project on kinship in Norway had also received a substantial grant from the Norwegian Research Council.[2] This grant enabled me to engage a research assistant, Kari Anne Ulfnes. Continuing the work of Esben Leifsen, who had been of great help in collating information of relevance to the project as well as giving practical support, she also took part in carrying out interviews. Her conscientious and dedicated work made it possible for me to combine my administrative and academic obligations at that time.

As the research proceeded, the project expanded far beyond its initial aims. First and foremost, my attention was drawn to the Norwegian legislation regulating assisted conception and the fact that this legislation differed from that of other Scandinavian countries and Europe more generally. Norwegian legislation was by far more restrictive. Thus, my curiosity was kindled: how to understand this reticence on the part of Norwegian legislators? My chance to explore this line of research was subsequently made possible by two larger collaborative research projects, which provided additional funding and, not least, the opportunity to engage with colleagues in Norway and elsewhere in stimulating and challenging intellectual exchanges. These projects were 'The Transnational Flow of Concepts and Substances', which ran from 2001 to 2004 and was funded by the Norwegian Research Council (NFR), and the 'Public Understanding of Genetics: A Cross-cultural and Ethnographic Study of the "New Genetics" and Social Identity (PUG)', which was funded by the European Commission and ran from 2002 to 2005.

That close collaboration with many colleagues over several years be a rich and rewarding experience is not a given. I was privileged. I want to thank my co-participants in the project 'Transnational Flow' at the Department of Social Anthropology, University of Oslo, for their interest and encouragement. Marianne Lien, who headed the project, and with whom I have collaborated closely on various occasions, took time to read an early version of this manuscript for which I am very grateful. She is a critical reader and her sharp comments have left their mark, but perhaps in ways she might not recognize. The Department of Anthropology has been supportive throughout, also providing a sabbatical at a critical time. PUG was an interdisciplinary effort including scholars from the UK, France, Spain, Lithuania, Hungary and Italy. My thanks go to all, but in particular to Jeanette Edwards for bringing us together and keeping us on track, and most importantly for her generosity and inspiration.

The year of 2005 I spent in Paris as Professor at the Centre de Coopération Franco–Norvegienne at the Maison des Sciences de L'Homme. This proved to be a very fruitful time as it also allowed me to present my work at different seminars and workshops. I especially want to thank Enric Porqueres i Gené whose unfailing enthusiasm and efforts to include me were very much appreciated. I also wish to thank the Groupe de travail d'anthropologie sociale comparative at the EHESS for their invitations to present my work at their seminar. The ensuing discussions were always both challenging and stimulating. In particular I want to thank Cécile Barraud, Jean Claude Galey and André Itéanu for sharing their thoughts and time with me. Ann Cadoret has been an especially thoughtful and inspiring interlocutor, prompting me to move in different directions. My colleagues at the Centre, Kirstin Skjelstad and Saphinaz Naguib, were both, in their different ways, always supportive of my work.

There are many people who at different times and in different capacities have directly and indirectly contributed to my thinking and work on this book. Most important are all those people who have given their time and their thoughts to an anthropologist and all her questions. I especially want to thank the Forening for ufrivillig barnløse (Association of the Involuntarily Childless, now known as the Ønskebarn) and their then leader Kaja Gruff Huster for taking an interest in my project and providing me with the initial introductions; to Mette Schmidt who got me going in Oslo; and to those women and men who shared with me their experiences of being involuntarily childless. I also want to thank Johan Hazekamp for his keen interest in my project in its early stages and for actively engaging me in his network. Berge Solberg has always provided helpful comments, and not least a quick response to my queries at a later date. There are also many 'experts' on different aspects of my research that have shared their knowledge and views with me and to whom I am grateful. These include medical doctors, heads of clinics, bioengineers, bioethicists, politicians, state bureaucrats and other specialized personnel. Other persons who have provided food for thought and encouragement at various points throughout these years are Eduardo Archetti, Joan Bestard, Ben Campbell, Geneviève Delaisi de Parseval, Olivia Harris, Keith Hart, Anne Hellum, Marcia Inhorn, Bruce Kapferer, Kjersti Larsen, Sarah Lund, Carles Salazar and Peter Wade. I have received excellent library support from Astrid Anderson. Cecilie Broch Knudsen suggested the cover image. Finally, I wish to acknowledge my two anonymous readers for their careful reading and pertinent comments, not all of which I have been able to incorporate. At Berghahn I wish to thank Ann Przyzycki DeVita and especially the in-house editor Charlotte Mosedale and the copy-editor Philip Thomas.

Oslo, 7 February 2012

Chapter 1

FRAMING THE ISSUES

> When the future is imagined through reproduction,
> visions are evoked of social relations and political order.
> —Ludmilla Jordanova, 'Interrogating the Concept of Reproduction'

> Some of the hold that biotechnology exercises over the imagination
> is its power to intervene in realities that already play a role
> in the way people think about themselves.
> —Marilyn Strathern, *Kinship, the Law and the Unexpected*

> In studying biotechnologies that transgress bodily, legal,
> philosophical, and spiritual boundaries, we must ask why in
> some locales such innovations raise little concern,
> whereas in others they create havoc.
> —Margaret Lock, *Twice Dead*

> Nature, as we know is neither stable nor fixed, and kinship is a
> key site in which distinctions between nature and culture,
> what is given and what is acquired, what is fixed and
> what is changeable, are produced.
> —Jeanette Edwards, 'The Matter of Kinship'

Introduction

The overall framing of this book is reproduction, writ large. It addresses some fundamental issues in contemporary Norwegian biopolitics. Reproduction is at the core of social transformation, and is central to both the material and symbolic understandings of social life. Reproduction – in its biological and social senses – is as Ginsburg and Rapp note, 'inextricably bound up with the production of culture' (Ginsburg and Rapp 1995a: 2). Thus, a study of social reproduction is also a study of cultural process. All societies have vested interests in reproduction, but, as anthropologists have documented, what these interests are and the way they are regulated – and not least perceived – locally vary. Spurred by developments in biomedicine, this book is an ethnographic account of locally embedded reproductive relations within a global world of reproductive possibilities. It is about procreation – and procreative practices – and how these are articulated at individual and social levels. More concretely it is about a specific trajectory of the politics of reproduction as these are translated into a legal framework, emphasizing the way this plays out on the individual and social body in Norwegian society. One important aspect of this process is the continual redrawing of the public–private divide, of state matters and personal matters. Another is the controversies surrounding the ascription of proper filiation, of 'making parents' (Thompson 2005); and yet another, the fear of eugenics. Moreover, the expansion of a global infertility market has come to represent a significant factor in various local perceptions of biopolitical practices. At the heart of these socio-cultural cum political processes are not only problems of conception but also, and more significantly, meanings of conception. These appear to be paramount to the biopolitical agenda in Norway.

The empirical focus of this book is the convergence of two phenomena: a new technology and a new law. The technology in question is reproductive technologies and the law that which regulates the use of biotechnology in medicine.[1] This book is about the making of that law (and its subsequent revisions); hence, it is about a legislative process and what the study of such a process may reveal about a particular socio-cultural reality.[2] At issue are some of the interfaces between technology, science and religion as these are articulated through law, and the concomitant legislative processes and practices. I am interested in the way these phenomena become intertwined, and in what these phenomena, when taken together, articulate with regards to fundamental values of relatedness and society in contemporary Norway. Hence, this book is also about kinship. I will explore these values by examining the significance of biogenetics for kinship relations, on the one hand, and a particular Norwegian notion of 'the sorting society' (*sorteringssamfunnet*), on the other. Moreover, I not only want to examine how these values are refracted in the legal discourse embedded in the law and projected by lawmakers but also how some of those subject to the law think, act and respond. By locating discrepancies as well as points of contention, I hope to disclose areas of moral indeterminacy and even controversy, as well as processes of naturalization.

In contrast to many countries throughout the world, where the different technologies of reproduction have been endorsed and even welcomed, Norwegian legislators have been much more wary. In fact, the Norwegian law regulating biotechnology has, until its recent revision in 2007, been one of the most restrictive in Europe, in terms of what it permits, and by implication what it prohibits.[3] In short, it prohibits egg donation and surrogacy; it prescribes the use of known-donor sperm; until 2007, it did not permit research on embryos; it limits the use and application of preimplantation and prenatal diagnosis (the law and the legislative process are detailed in Chapter 3). This law is based on a precautionary principle: better safe than sorry.

The first Norwegian law addressing assisted conception was passed in 1987.[4] It came in the wake of the birth of the first so-called 'test-tube baby' (*prøverørsbarn*) or IVF child in Norway in 1984. At that point, legislation was deemed necessary, lest matters get out of hand (Melhuus 2005). However, the passing of the 1987 law – as well as subsequent legislation in 1994, 2003 and 2007 (to which I return in Chapter 3) – was also an explicit response to a transnational phenomenon: the overall accessibility of knowledge and technologies that can be made to act on human reproduction. Nevertheless, it does not follow from a general agreement about the need to regulate that the law necessarily be restrictive,[5] at least when compared with similar legislation in other countries.[6] With the exception of Germany and Italy, most other European countries have legislation that is more permissive than Norway, including the other Scandinavian countries. It is this fact – and what it entails for all those concerned – that has spurred my curiosity and which singles itself out for ethnographic attention. This law limits people's choice as to how they wish to procreate within the borders of the nation-state, while at the same time prompting people to travel abroad in order to obtain treatments not permitted in Norway. Hence, this legislative act is instrumental in what has been coined as 'reproductive' or 'fertility tourism' (see Deech 2003; Storrow 2005).

The law, then, reflects a discrepancy between a legal norm and actual practices. Thus, my queries are directed towards questions of moral ambiguity and moral dispute, and how significant moral values, both implicit and explicit, are variously articulated and, not least, contested. The examination of a legislative process and codification of a law within a broad context suits this purpose well. Laws are significant because they both reflect dominant social concerns and values and are normative, in the sense that they seek to regulate and/or improve current practices. Moreover, because laws are cumulative, it is possible to trace changes in the cultural climate. Furthermore, because laws are explicit and unequivocal in their formulations, it is possible to identify the moral principles that are evoked and challenged (Melhuus and Howell 2009).

My main focus is on reproductive technologies and various forms of assisted conception. Hence, within the broad category of biotechnology, I am specifically concerned with what has been termed new reproductive technologies (NRTs), which, not being so new any longer, are now more often referred to as assisted reproductive technology (ART), where in vitro fertilization (IVF) is central. However, I also include artificial insemination by donor (AID), which strictly speaking does not fall into the category of reproductive technology,[7] but is part of the routine practices of assisted conception (and regulated by the legislation), as well as other technologies associated with ART, such as prenatal diagnosis (PND) and preimplantation genetic diagnosis (PGD). Although these latter technologies are not necessarily implicated in acts of assisted conception, they are definitely part of technologies of reproduction and, hence, contemporary procreative practices. In Norway, moreover, these technologies are also regulated by the Biotechnology Act.[8]

Although the nexus around which my arguments revolve is the Norwegian Biotechnology Act, I am as much interested in what the law produces – that is, its effects – as I am in what has produced the law. I am as interested in context as I am in text. Hence, I cast my net broadly to include those people, events, processes (historical as well as contemporary), discourses, imaginations (not least imaginations) and representations in order to delineate what I call a procreative universe – that is, a socio-cultural space which in some way embraces the complex reality in which biotechnologies and assisted conception are embedded.[9] In addition to notions of kinship and relatedness, nature and nurture, of the individual and choice, this includes ideas of society and the state, attitudes to science, and not least what expert knowledge 'means'. It also implies drawing attention to the notion of borders, be they moral or territorial, and acts of transgression. Underpinning these processes there is on the one hand a tension between biology and kinship provoked through an identity discourse that subsumes specific notions of biogenetic relatedness and notions of rights. On the other hand, these processes also reveal a tension between individual and society, which is most explicitly evident in the rhetoric surrounding the 'sorting society' (see Chapter 5).

Although this book, then, is about reproductive technologies and what they propel in the way of significant articulation – for example in the form of law or in meanings of relatedness – this book is also about Norway, or more precisely about certain aspects which contribute towards an understanding of some fundamental dimensions in contemporary Norwegian society.

Kinship: A New Beginning?

The birth of Louise Brown in 1978 represented a major breakthrough in reproductive medicine. She was the first child to be born as the result of a new technique, in vitro fertilization (IVF), making conception outside the body possible (Edwards and Steptoe 1980). This extraordinary achievement was recognized in 2010, when Robert Edwards, who pioneered IVF and was responsible for the birth of Louise Brown, was awarded the Nobel Prize in medicine. Briefly, IVF involves extracting eggs from a woman's ovaries and combining them with sperm in a laboratory dish in order to achieve fertilization. The resulting embryo(s) are then transferred into the uterus of the woman. IVF not only allows for fertilization to occur without sexual relations, it also, and perhaps more radically, allows for the movement of eggs between women (egg donation) and various combinations of sperm and egg donors, blurring the categories of maternity and paternity and creating potentially multiple 'parents'.

This event was to have repercussions worldwide, as the knowledge and technology became internationally available, albeit not necessarily accessible. Accessibility in this case is first and foremost a question of cultural acceptance – of legitimacy, not of economy. Technologies are not innocent and they do not travel unhindered. As Inhorn (2003) demonstrates for the case of Egypt, reproductive technologies are not transferred into cultural voids. On the contrary. They are domesticated, 'shaped by local social, cultural, religious, and scientific traditions' (Inhorn 2003: 3) – and, I would add, legislation (or the lack of it). The various local appropriations and incorporations of reproductive technologies that followed in the wake of Louise Brown have set a new agenda, contributing towards a certain reordering of relations in the world (Ginsburg and Rapp 1995a; Melhuus 2007). Questions of infertility and reproductive desires are now an issue, not only for medical personnel and those infertile people (couples or single individuals) who wish to put those technologies to use, but also feminists, social scientists, bioethicists, the public at large, even film-makers, as well as religious communities and nation-states, all of whom have been drawn into and engaged in debates about the potential use and abuse of reproductive technologies. Imaginations have been triggered.

The birth of Louise Brown came at a time when family formation was already undergoing significant changes in Europe and North America. Divorce rates were increasing and divorce was, in many circles, morally accepted, as were (almost by implication) reconstituted families; single parenthood was less stigmatized (implying that the stigma attached to children born out of wedlock was waning); and homosexual relations were gaining acceptance. Although there was an increased interest (among social scientists and others) in these transformative processes, the advent of the new reproductive technologies rekindled and redirected the thrust of kinship studies.[10] These technologies undermined what was perceived as the natural basis for procreative practice, and thereby challenged some of the most fundamental tenets of procreative ideologies, upsetting 'normal' ideas of family formation. This has had many implications, which the burgeoning literature on this topic since the mid 1980s bears witness to.[11] One significant aspect is that the focus of attention (in kinship studies) shifted from (predominantly) small-scale non-Western societies to Euro-American ones.[12] Moreover, in order to grasp these new forms of making kin it was necessary to engage with a series of phenomena not usually understood as relevant to – or implicated in – kinship: advanced technologies, scientific knowledge, and detached bodily substances (such as sperm, eggs, embryos); the experts, practitioners, donors – all had to be contended with. Both the objects and the subjects of kinship are (in a certain sense) of a different kind than that which prevailed earlier. In other words, the practices of procreation that have come in the wake of ART involve a whole new repertoire, new sets of relations, as the kin-making group expanded to include all the people, elements and substances – and the relationships between them – involved in the creation of a baby. It is as Thompson suggests: ARTs demand as much social as technological innovation to make sense of the biological and social relationships that ARTs forge and deny (Thompson 2005: 5). Over the past decades, considerable effort has been put into making some sense of the link between these biological and social innovations among scholars and lay people alike. Much effort has also been made to grasp how different subjects and/or publics (Edwards 2009) make that link. That has been one of the major contributions of anthropologists.

Among many other things, this shift in the drift of the anthropological gaze has implied a reconceptualization of 'the field', both regarding the type of localities appropriate for the study of kinship practices – for example, fertility clinics – and their meanings and the extension, or remaking, of the field to include many non-localized sites. Multi-sited fieldwork (Marcus 1998) is the rule rather than the exception in rehearsing contemporary European and North American notions of relatedness and belonging. Not only, then, do the new kinship studies construct new sites for exploring these themes, but also the kinds of ethnography that are generated with regards to the interrelated issues of biotechnology and kinship are varied and cover a broad range.[13] These ethnographies also beg a different set of questions, being concerned with and deeply implicated in issues of knowledge, science, and to a certain degree ethics, and how these can be brought to bear on notions of personhood, identity, family and kinship, and vice versa.

This rekindling of kinship studies represents an exceptional opportunity for comparative work. A careful examining of how kinship is conceived, what constitutes significant relatedness, and what meanings are drawn from – or attached to – the technologies as well as the substances involved (such as egg and sperm) will reveal differences as well as similarities. One area of evident difference is precisely that of legislation and the types of legislation that have (or have not) been put in place. Legislation and policy is, as Edwards notes, 'a site where certain kinship understandings are crystallised' (Edwards 2009: 7). Thus, the national contexts within which biotechnologies operate are significant, and government one important site for ethnographic enquiry in order to tease out some of the substantive contrasts. Biopolitics takes many different forms, and an exploration of such politics allows for a systematic comparison across Europe while also bringing to the fore some profound differences between the USA and Europe,[14] as well as other parts of the world. The present ethnography will hopefully also serve that purpose.

Some Other Issues

I stated above that the kinship studies generated as a result of reproductive technologies implied a shift not only in analytical perspectives but also as to where the research was carried out and the kinds of sites that were constructed as appropriate arenas. Much of the earlier work that laid the ground for subsequent research was based in Europe and the USA. This has to do with where the reproductive technologies were first put to work – and significantly how they were inscribed in various public discourses. Reproductive technologies have (and had) an extremely high impact factor (to use contemporary jargon), mobilizing publics of many different kinds. The

Warnock Committee's report to the British Parliament (Warnock Committee 1985) and the case of Baby M are but two well known and early examples of how the uses (potential and actual) of assisted conception and reproductive technologies were debated and publicized, initiating as well as underpinning subsequent discourses on reproductive technologies. In both cases, anthropologists were quick to see the implications of these events for the understanding of kinship relations (see Rivière 1985; Dolgin 1992; Fox 1997 [1993]; Franklin 1999 [1993]). Whereas the Warnock Committee report was commissioned in 1982 by the British government to address, evaluate and suggest proposals for the application of various forms of assisted conception (see Shore 1992), the case of Baby M is an actual case of surrogacy, tried in the US court system in 1987, contesting parental rights to a child born by a surrogate mother (Dolgin 1997). Although these two examples are drawn from very different contexts and represent quite different kinds of processes (one an act of government, the other a case of litigation), they nevertheless indicate the kinds of issues that were – and to some degree still are – at stake. These issues range from the fragmentation of motherhood, questions of reproductive intent and the privileging of biological connectedness, and the ensuing problems of ascribing proper filiation to the child – for example, a baby born with the use of donor eggs and surrogate mother – to the moral status of the embryo and its use (or not) for the purposes of research. Not least, the two cases indicate a fundamental difference in attitudes to the question of the commercialization of bodies and body parts, distinguishing the USA from most European countries. This in turn reflects basic differences regarding the role of government, understandings of the divide between private and public, as well as the relation between individual, family and society.

Socio-cultural studies of assisted conception and reproductive technologies have since been carried out in many different parts of the world (e.g., Kahn 2002; Inhorn 2003; Simpson 2004; Shimazono 2005; Roberts 2007). Grounded as these studies are in other realities, they yield other insights while drawing attention to yet other issues. Prominent among these is the role of religion and religious beliefs and doctrines in the way reproductive technologies are deployed in different contexts. As anthropologists (and/or other scholars) explore the incorporation of reproductive technologies in Jewish or Muslim societies, a whole other set of values and considerations are drawn into the framework. Stressing the importance of religiously based moral systems in conjunction with ARTs, Inhorn (2007) points to the differences between Sunni and Shi'a Islam in coming to terms with assisted conception and reproductive technologies. In this connection, she suggests that Iran is 'the country to watch, as it has been on the "cutting edge" of new reproductive technologies' (Inhorn 2007: 195).

Within predominantly secular Christian societies – such as Europe and the USA – there is (in contrast to, for example, Jewish and Muslim societies) an exceptional preoccupation with the status of the human embryo. The kinds of arguments – *pro et con* – that are brought to bear on what status to grant the embryo and which ones (in any one situation) carry most weight have far reaching consequences for the kinds of relations and/or exchanges that the embryo may enter into. In those contexts where the embryo represents a singular object (Kopytoff 1986), the points of resistance – that which holds the embryo back so to speak – are of paramount concern. As we shall see, in the Norwegian case, there are different factors at work to immobilize the embryo, thereby restricting its potential circulation and the uses and/or practices to which an embryo can be submitted (for example, surrogacy or stem-cell research). Important among these are Christian notions which implicitly and explicitly inform the way the embryo is understood – and thus also what ends it can (or cannot) serve. The moral status of the embryo is a nexus around which much debate has turned, a focal point being whether to grant the embryo absolute human dignity. In other words, the discussions have centred on the distinction between human life and human being and whether the moment of conception confers complete humanity on the embryo.[15]

Law/Imagination

It is my contention that an examination of laws and legislative processes can contribute towards a better grasp of significant contemporary socio-cultural processes. Law can be viewed as a system of cultural meanings where practices of law (which in this case includes legislative processes) are 'productive of meaning' (Merry 1992: 361). This perspective is in line with Geertz's (1983) approach, from which I take some inspiration, although his agenda is somewhat different to mine.[16]

In his essay 'Local Knowledge: Fact and Law in Comparative Perspective', Geertz argues that 'law is local knowledge; not just as to place, time, class, and variety of issue, but also as to accent – vernacular characterization of what happens connected to vernacular imagination of what can' (Geertz 1983: 215). Geertz insists that law be understood as a local phenomenon and argues forcefully for the significance of context, stating that 'the cultural contextualization of incident is a critical aspect of legal analysis, there, here or anywhere, as it is of political, aesthetic, historical, or sociological analysis' (Geertz 1983: 181). Thus, he anchors the study of law firmly in time and place, arguing that '[law] is constructive of social life, not reflective, or anyway not just reflective, of it' (Geertz 1983: 218). Geertz is interested in processes of adjudication, and more generally in the relationship between fact and law, reflecting on the place of fact in a world of judgement. He argues his case through the comparison of three Indonesian systems of law (Islamic, Indic and *adat*). My focus is a concrete process of legislation (and not concrete litigation processes or dispute settlements),[17] and how that process works so as to establish certain undisputable facts. This will necessarily demand both a broad and detailed contextualization.

The incident, to put it in Geertzian terms, that concerns me is the making of a law: its codification. This is of course not the same as a legal incident involving a court case and a settlement. Yet as I hope to show, there is a process of adjudication of sorts, which culminates in Parliament's passing of a law. This process takes a very different form (than a case of litigation), not least because of the time dimension: it takes years. And in the course of those years changes occur. The process involves a trail of official documents (such as commissioned reports, white papers, hearings); committee meetings and debates in Parliament; extensive media coverage; as well as various types of public hearings. It involves a whole range of actors, *inter alia* the involuntary childless, medical personnel, bioengineers, bureaucrats, politicians, bioethicists, researchers and activists. In short: a procreative universe, the limits of which are hard to determine. However, one of the most salient features of this legislative process (and in stark contrast to a court case) is its indeterminacy, in the sense that it is never completely final, always subject to revision. It moves along with the times. This has as much to do with the continual production of new knowledge and new technologies, giving rise to new medical – and ethical – scenarios as it has to do with ever shifting political constellations. It is also a result of people's – that is, those subject to the law – actions. What people do influences legal practice. Actions have effects.

There is, as Rabinow says, commenting on the French cultural milieu but in terms which are applicable to Norway, 'a profound uneasiness about the consequences of recent technological and scientific invention and discoveries' (Rabinow 1999: 17). Biopolitics reflects this state of affairs. In trying to come to grips with the uncertainties that biotechnology are seen to entail, law becomes a principle technology of government. Moreover, the law articulates legislators' efforts to make explicit sense of a world that is at one and the same time both real (in the sense of actual) and imagined. This 'sense' does not necessarily mirror the 'sense' of some of those subject to the law (as we shall see). Whereas legislators aim to govern the imagined risks of biotechnology by applying a precautionary principle (Pottage 2007: 333), people (in need of such technological assistance) imagine biotechnology as a means to another end: a potential to be realized in their efforts to become a family.

There is no doubt that imagination plays – and has played – a significant role in conceptualizing the potentialities of biotechnology (Melhuus 2007). What is striking is the range of possible futures that such technologies provoke, the antagonistic positions created. Imagination is a social practice with significant ramifications (Appadurai 2000). Needless to say, the relationship between what the imagination produces and what produces the imagination is a dialectical one. Socially relevant imaginations are limited and contextually bound, and the relevant contexts for stretching the imagination shift, creating new topographies and new horizons. These topographies vary with underlying perspectives, which are complex compositions involving morality, emotions and intentions – as well as incidents elsewhere and the new technologies.

Imagination is not coincidental and works in many directions, serving both as a break and an accelerator on socio-cultural processes.

In his essay on law, Geertz says that '"law", here, there, or anywhere is part of a distinctive manner of imagining the real' (Geertz 1983: 184). He suggests that law be conceived 'as a species of social imagination' (Geertz 1983: 232), and that attention should be focused 'on how institutions of law translate between a language of imagination and one of decision and form thereby a determinate sense of justice' (Geertz 1983: 174). However, this has to be done in context, in detail, and with a stress on local knowledge (of objects and people), meanings and values. Only then will law disclose a connection between what happens and an imagination about what can happen. This is, in my understanding, what Geertz refers to as legal sensibility: 'this complex of characterizations and imaginings, stories about events cast in imagery about principles' (Geertz 1983: 215). Although the notion of legal sensibility is not so important to my discussion, the ideas captured by this concept are. It is the relations between events/practices (such as the making of IVF babies) and principles (such as that which underwrites the prohibition of egg donation), between substances (such as embryos) and their classification that interests me. Imaginations are embedded in these processes. Following this train of thought, it is possible to view the law regulating assisted conception in Norway as a product of the imagination, a cultural artefact, but one that is implicated in 'legal fabrication' (Pottage 2004). I will explore what constitutes and counteracts that imagination; what grounds the figure projected through the law. That ground has to do with ideas about what is natural, notions of the family and of maternity and paternity – in effect: kinship. Thus, that effort will necessarily take me beyond the law as such to those who are affected by it, such as the involuntary childless.

The Involuntary Childless

I have approached the study of reproductive technologies, law and kinship in Norway through the involuntary childless (*ufrivillig barnløse*). They represent an exceptionally fruitful point of entry in order to understand contemporary kinship relations, through their efforts at family formation and their reflections about meanings of relatedness, in coming to terms with their situation. The term 'involuntary childless' is the indigenous term (ascribed and self-ascribed) used to denote those who wish to have children but are unable to conceive.[18] The involuntary childless are people, normally couples, who are driven by their desire to become a family, to have a child they can call their own: 'an own child' (*et eget barn*) as they would say. The two alternatives available to realize this desire are either assisted conception or adoption. As very few local children are put up for adoption in Norway, adoption will invariably be transnational (Howell 2006). According to Howell, there were more than 17,000 transnationally adopted individuals living in Norway in 2004, and about 700 arriving each year (Howell 2006: 25). Many of those who opt for adoption have first been through unsuccessful cycles of IVF. Thus, the two practices are social phenomena that belong to the same order and are mutually implicated forms of assisted procreation. Nevertheless, in some senses and some contexts, adoption, paradoxically, plays the role as a 'natural' model (of kinship) for the various forms of assisted conception (Melhuus and Howell 2009). In contrast to reproductive technologies, which have opened up a whole new world of possibilities, adoption is a practice that is familiar; it is known, and has been for a long time. In addition to the concrete experiences of adoption, this knowledge is also lodged in legislative processes.

Adoption in Norway has a long legal history. The first adoption law was passed in 1917, and adoption belongs to the corpus of laws dealing with children and family.[19] It is therefore not surprising that this existing knowledge about the incorporation of non-biological children into the bosom of the family is mobilized to make sense of babies created as a result of ART. The point of interest is how this is done (see Haimes 1998). After all, 'having' a baby by adoption or by assisted conception are two very different ways of procreating; they are also two very different ways of becoming someone's child. The analogy works by making some aspects of adoption explicitly relevant, while others are silenced as a consequence. One of the aims of this book is to disclose such cultural mechanisms. As I will show, two phenomena are central in this exercise: the notion of rights and the significance of biological origin for identity.

Harris observes that 'anthropology characteristically chooses as its field of study those who are at the frontiers of legality, and anthropologists seem to have an affinity for those whose relationship with law is at best ambivalent' (Harris 1996: 3). I am sure she did not have the involuntary childless in mind. Yet, many involuntary childless in Norway do have a very ambivalent, if not outright negative, attitude to the law. This is accentuated because the law puts them in 'difficult moral predicaments by the fact that [it] forbids, or even criminalizes, actions which for the people concerned are acceptable or desirable within their own moral code' (Harris 1996: 2). Despite the fact that the law actually grants infertile couples access to infertility treatment (within certain limits) through the public health service (implying that most of the costs are covered) for many involuntary childless, the law through its proscriptions is an impediment to their greatest desire: to have a child they can call their own. Moreover, many of them in fact undermine the intents of the law and travel abroad for treatments not permitted in Norway, entering a circuit of 'reproductive tourism'. They are well aware that a pregnancy cannot be stopped at the border and, in the case of surrogacy, well informed about how to enter a child born abroad to the Kingdom of Norway (see Chapters 2 and 4).

In coming to terms with their infertility, the involuntary childless have to make some sense of their situation. They have to mobilize different images in order to be able to make a decision, a decision that will in one way or another determine their life trajectory. However, the decision-making process is just that – a process which is made up of many decisions, some major and some minor. Most of the involuntary childless I spoke to were in the middle of that process. They were reflecting on their options and attempting to make informed decisions. They were trying to make sense out of a situation that for many appeared meaningless. Some were already 'in treatment' as they would say; some were hesitating between adoption and infertility treatment. Some had concluded the three cycles granted within the Norwegian health-care system and wondering whether they should 'go private', adopt or give up. Others were considering whether they should go abroad for treatments not accepted in Norway. What they all had in common was a firm conviction that to remain childless was not an option. They were also exceptionally reflexive about their situation. They were extremely self-conscious about what being a family meant to them and the efforts they were willing to make in order to become one. Underpinning their deliberations was what having an 'own child' means. This entailed imagining the different forms of relatedness that are implicated in the various procedures available – here (in Norway) or elsewhere. Their attempts to convey the significance of this term ('own child') have also informed my understanding of kinship in contemporary Norway, so much so that I devote a whole chapter to this theme (Chapter 2).

Some Reflections on the Precautionary Principle, and a Bit More

The Norwegian Biotechnology Act of 1994 (and its subsequent revisions) is based on a precautionary principle.[20] This is explicitly stated and also implicitly evident in the restrictions that the law encodes. Thus a few words on this principle seem pertinent. I stated above that the law articulates the legislators' efforts to make sense of a world that is both real and imagined, and that social imagination is embedded in the processes that propel the law. The application of the precautionary principle is an expression of a particular imagination; perhaps even an act of the imagination. On the one hand, this imagination translates biotechnology not only into a legal framework but also into a precautionary mode. On the other hand, the precautionary principle translates between the language of imagination and the language of decision (to paraphrase Geertz). It is imagining what these technologies might entail (in the future) that informs a restrictive policy. It is the corrosive effect of the technologies on core values in Norwegian society that guides this 'regime of caution' (Pottage 2007: 334).

In his broad review of the socio-legal implications of biotechnology,[21] Pottage (2007) also discusses the precautionary principle. He directs attention to what he terms 'modes of bioreflexivity' tied to notions of risk and uncertainty. He contrasts the speculative mode where uncertainty – or risk – is capitalized, with a mode where uncertainty is 'accentuated and dramatised as a political predicament' (Pottage 2007: 333). Whereas the speculative mode is based on an alliance between investors and consumers with regard to the promissory futures of biotechnologies,[22] the regime of caution rests on the precautionary principle. The essential premise of this principle, Pottage states, is that 'technologies that have the potential to cause massive and irreversible harm should be subject to regulation even in the absence of clear evidence of the likelihood of such harm ... [T]he precautionary principle effectively becomes the basis for a very specific mode of balancing technological evolution against potential social harm' (Pottage 2007: 333).

Among other things, it is the content of this balancing act that I try to trace empirically by exploring the legislative process regulating biotechnology in Norway. In this case, the precautionary principle is not just a legal strategy – or policy – for managing risk, it is also a political strategy for protecting and promulgating socio-cultural values deemed central to Norwegian society. In so far as an idea of the integrity of the body is the locus of biotechnological interventions and hence is mobilized as a biopolitical device (Pottage 2007: 333), I would nevertheless argue that what is 'at risk' in Norway is not so much the individual body but the social body. The application of biotechnologies in medicine is seen not only as potentially upsetting fundamental kinship categories but also (and perhaps more importantly) they are also seen as challenging the very idea of society, or rather some fundamental principles that constitute society. However, and this is important to keep in mind, the incorporation of reproductive technologies in medical practice in Norway is – and has been – contested. There is no overall consensus on these policy matters, although the precautionary principle as such is not controversial. Moreover, as the Biotechnology Act is yet again (in 2011) under evaluation, it is reasonable to assume that the law will be revised.[23] Thus the processual aspects of biopolitics will continue both as a response to the very developments in biotechnology and as a response to the political climate in Norway.

The fact that the Norwegian law regulating the use of biotechnology is restrictive was not the only reason – or reason enough – to make it my main concern, although exploring the reasons for why a precautionary principle seemed prudent is culturally revealing. Rather, what draws my attention is the discrepancy between what the law permits and what many of those subject to the law (practitioners and patients alike) want. Thus, I am interested in a double movement that takes into account the types of subjects the law produces as well as the types of subjects that have produced the law. What becomes evident through this double movement are the tensions this relationship generates, making visible a socio-cultural disparity. I explore this area of moral dissonance that the law throws into relief. How is this discrepancy articulated and how can it be understood?

First and foremost, it is the plight of the involuntary childless themselves that comes to the fore. They were concerned with the possibilities they had of becoming parents: of having a child, of if and how the law impeded their choices and what they were to do about that. Their concerns and their evaluations of possible options (within and outside Norway) reveal significant insights as to the meaning of kinship relations, in particular with regards to the nature/nurture configuration and the shifting articulation of values placed on one at the expense of the other. Secondly, there have been many and heated public debates about assisted conception, technologies of reproduction and the law. Ever since the first attempts were made to legislate artificial insemination by donor in the 1950s, these issues have been on the public agenda. The media coverage has at times been overwhelming. These media events occur most often in conjunction, but by no means only, with legislative processes – be it the introduction of a new legislation or its subsequent revision. Time and again issues pertaining to procreative practices will reach the national – and even international – headlines. To give but one example: in October 2004 one of Norway's major tabloid newspapers ran the following headline: 'King Haakon was not [King] Olav's father?' The subtext reads: 'In his new book on the monarchy, biographer Tor Boman Larsen suggests that Sir Francis Laking, the king's personal doctor, is King Olav's biological father'.[24] The following day, several newspapers ran a follow up and the item even merited international attention. The UK's *Daily Mail* had the following headline: 'Was a King of Norway really made in England?'[25] Meanwhile, *The Times* ran an article entitled 'The British Royals of Norway'. The opening sentence reads: 'A British baronet secretly sired the late King of Norway using a primitive form of artificial insemination, according to a new biography that casts doubt on the Norwegian royal bloodline'.[26]

Granted, this is an unusual story, and matters relating to the royal family invariably attract media attention and often sensational headlines. The story not only deals with sperm donation carried out in 1902, evoking ideas of illicit sexual relations, but also challenges the royal imperative of impeccable descent lines predicated upon biological connectedness, while at the same time introducing the uncertainty of paternity. If headlines are anything to go by, the news was received as sensational. Nevertheless, perhaps the most sensational aspect of this piece of information is that the possible conception by donor sperm of the former king of Norway has not made a difference to the status of the royal family. The story did not evoke strong moral or emotional reactions among the public at large.[27] This may tell us something about attitudes to the royal family, but I believe, in the context of my own research, it tells us more about perceptions of kin relations and a willingness to grant kin-status as a result of social intimacy. It also tells us something about changing values. The media's continual preoccupation with procreative events says several things: that there is a pervasive public concern, that these issues draw public interest, and not least that it is considered legitimate to introduce such intimate questions to the Norwegian *offentlighet* (public domain), thereby redrawing the line between private and public affairs.

Finally, pursuing a legislative process allowed me to incorporate a temporal dimension and hence the possibility of examining continuities and change. At a more concrete level, this historical perspective has two dimensions. I start my explorations into this field with the first attempts to legislate artificial insemination by donor (AID) in the 1950s and continue down to the revision of the biotechnology law in 2007 (Chapter 3). Thus I trace the development of a specific biopolitics. At a later point in the book (Chapter 5), in conjunction with my discussion of the notion of the sorting society, I also look back – to the 1930s – drawing attention to eugenics, the question of racial hygiene and the sterilization law (passed in 1934) and its relevance for policies today. Lending some historical depth to these issues makes evident the socio-cultural changes that have taken place over the past fifty years. Perhaps the most significant of these changes (with regard to procreative practices) has been the shift in preoccupations about sex without procreation (contraception, abortion) to procreation without sex (AID, IVF, surrogacy). These changes have in turn had implications for the way that reproductive technologies have been received and incorporated in Norwegian society. Overcoming infertility through the use of reproductive technologies is today culturally accepted in Norway. This is in itself an important datum. The very change in vocabulary from artificial conception (*kunstig befruktning*) to assisted conception (*assistert befruktning*) is indicative of this ideological shift, pointing to the routine incorporation and naturalization of reproductive technologies. No longer artificial, they have been reclassified as 'assisting' nature (see Chapter 6). Thus, a historical perspective allows for a simultaneous focus on both continuity and change, allowing for a better insight into the values being challenged and transformed. It also provides a contextualization and hence better understanding of the overall thrust of emergent dominant discourses.

In Chapter 4 I focus specifically on two discourses: the debates surrounding anonymous sperm donation and the prohibition of egg donation. Whereas Norway rescinded donor anonymity in 2003, egg donation has been consistently forbidden. Recognizing the difference between these forms of assisted conception, I examine the way these discourses, nevertheless, speak to some of the same issues. I analyse these two phenomena in terms of motherhood, fatherhood and questions pertaining to filiation. To anticipate an argument: in Norway (within the terms set by the present law) it appears that paternity is becoming more like maternity, rather than the other way round. In this process, biology is viewed as constitutive information, in the sense that Strathern talks about it: 'some kinds of information come with built in effects' (Strathern 1999: 65; see also Pottage 2007).[28] Applying a biological principle, the tendency is to biologize notions of identity, implying that knowledge of biogenetic origin is seen as fundamental to knowing who you are, in the sense that a part (gene) comes to stand for the whole (person).[29] The role of biology (read broadly) as constitutive information is also apparent in the debates about the application of preimplantation diagnosis (PGD) and prenatal diagnosis (PND). However, in this context it takes a different form: here it is a question of what (and when) information (about the condition of a fetus or embryo) is to be revealed to a pregnant woman. This issue arises in conjunction with restrictive proposals regulating the application of PGD and PND. It is through these debates that a fear of eugenics and selective abortion is voiced and articulated through the notion of the 'sorting society'. In Chapter 5 I focus on the persuasive power of this rhetoric, exploring the way these technologies are framed. In order to substantiate my argument I incorporate some reflections on Norway's history of racial hygiene. I shift my perspective from problems of relatedness (as demonstrated through filiation) to the relationship between individual and society with a focus on knowledge and choice. I demonstrate how arguments for a restrictive legislation are translated and enacted through an ethical publicity that privileges society over individual autonomy.

The State of Norway and the Notion of Equality

At the start of this chapter I intimated that this book is not only – or just – about reproductive technologies and law and kinship, but also a contribution towards an understanding of contemporary Norwegian society. However, in order to substantiate such a claim, I need, by way of conclusion, to add another frame that will bring other important dimensions to the fore, while serving as a backdrop against which the ethnography that follows should be read. I make note of the 'state' of Norway and discuss the Norwegian notion of equality. The former has to do with the values attributed to the welfare state and the role government plays, and the latter has to do with the notion of *likhet* (equality) and its status as a gate-keeping concept in the anthropological approach to studies of Norway (Vike, Lidén and Lien 2001: 12). These two are mutually implicated. As this book deals with legislation – a matter of state – and, moreover, illustrates particular articulations of equality with regard to values of relatedness (such as in Chapter 2), it seems pertinent to present these issues within a broader framework, albeit briefly.

As the historian Dahl points out, in his discussion of the historical roots of Nordic social democratic ideology: 'The Nordic equity ethos ... appears to apply both to the political action of levelling out – making the rich pay, taxing the top – and, in a jealous comparison, of making sure that nobody overtakes and passes you in position or possessions' (Dahl 1986: 99).[30] Although this statement is a simplification, it nevertheless crudely captures the double meaning with which notions of equality are imbued. They guide matters of state as well as social interaction. But they do so in different ways. It is, therefore – as Vike, Lidén and Lien suggest – useful to distinguish between 'equality as a premise for social interaction and equality as a regulating principle for the welfare state' (Vike, Lidén and Lien 2001: 16).[31] However, in practice these two meanings of equality are often conflated, where one is used to serve the purposes of the other (Lien and Melhuus 2009). Thus, as cultural categories they can be hard to disentangle. And it is, perhaps, precisely this entanglement that lends the notion of equality its rhetorical power in Norwegian society.

For Dahl it is the Norwegian term *rettferd* (connoting fairness, justice) that best captures what he denotes as an 'equity ethos', covering both the distributive and the equilibristic connotations of the concept. Drawing on Gullestad's work,[32] Vike, Lidén and Lien address the complex and multifaceted expressions of equality in Norway and seek to explore an established assumption that people in Norway are more concerned with *likhet* than people elsewhere, and that *likhet* is an especially important value (Vike, Lidén and Lien 2001: 12). The Norwegian notion of *likhet* has different connotations. As Gullestad has pointed out, *likhet* is understood both as equality – in the sense of being of equal worth (also denoted as *likeverd*) – and as sameness, being of the same kind or culturally similar (Gullestad 1989, 2001). There is a qualitative difference between these two meanings, but this is rarely made explicit. Gullestad shows how establishing 'sameness' in social interaction draws simultaneously on 'being of the same kind' and on 'being of equal worth'. Thus, the concepts of social equality and cultural similarity are easily conflated.

The Norwegian notion of equality is ambiguous and this ambiguity explains the paradoxes contained in the concept's rhetorical power (Vike, Lidén and Lien 2001: 16). Gullestad suggests that there is a discursive – or interaction – style 'in which commonalities are emphasised, while differences are played down' (Gullestad 2002: 47). She calls this style 'imagined sameness' (Gullestad 2002: 47). It is not difficult to see the implications of this style for establishing consensus. This would apply not only to situations of face-to-face interaction but also to political processes.[33] The problem, of course, is what such cultural mechanisms conceal of recognized differences, hierarchies or even outright antagonisms. What both the terms *rettferd* and *likhet* share is their multivocality and their underlying appeal to a higher order value of equality. The Norwegian notion of *likhet* can, as *rettferd*, also be inscribed in an equity ethos. This ethos not only informs the perceptions of the state but also informs state practice – for example, in legislating assisted conception and in the debates surrounding equal access to infertility treatments.

Government in Norway is based on the principles of the welfare state. Norway has a system of social welfare provision that includes free health care, free public education even at university level and a compulsory superannuation savings programme. Social welfare is understood as provided to all citizens (there should be no differential treatment) and is in principle not based on any criteria of need.[34] The edited volume *The Social Construction of Norden* (Sørensen and Stråth 1997b) provides a broad historical introduction to Nordic cultures and values and seeks to cast light on the particular path to modernity that characterizes the Nordic welfare states. As the title indicates, the volume addresses cultural templates, drawing attention to significant processes, values and tensions and their various articulations. Although aware that there are substantive differences between (and I assume also within) the Nordic countries, the editors nevertheless argue that a 'unique but "different" political culture emerged in the Nordic countries', and that 'the tension between freedom and equality ... was better contained than elsewhere' (Sørensen and Stråth 1997a: 3). Moreover they stress that '[the] specific Nordic Protestant ethic was a cultural genre which preserved a principle that was not only different from the Catholic cultures but openly hostile to them – the principle of unconditional personal freedom and the supreme value of the individual' (Sørensen and Stråth 1997a: 4). Their main concern is to question whether there exists a specific Nordic model of social organization with equality and welfare as its distinguishing marks. And their answer, not surprisingly, is that this model – as an identity producing projection from within and outside – was constructed 'in order to cope with specific political requirements. One of the key elements in this construction was history' (Sørensen and Stråth 1997a: 21). The volume collects different takes on this history.[35]

A core principle of the Nordic value system according to Stenius is that of universality (Stenius 1997: 169).[36] 'This principle', he says, 'gains substance at both the social and the political level, and its ideological prerequisite is that the state is seen as good' (Stenius 1997: 170). In the same vein, Sørensen and Stråth argue that the image of the state is one of a potential instrument of reform rather than repression (Sørensen and Stråth 1997a: 7), whereas Trägårdh posits that 'the central organising principle of the Nordic welfare state ... is the alliance between the state and the individual' (Trägårdh 1997: 263). What these statements indicate – or perhaps even presume – is that there is an abstract notion of trust between the state and the individual, and that this trust is part of the social contract. Such trust is not easy to document empirically, but Stenius claims that there is a 'trust in societal solutions' (Stenius 1997: 171). One way such solutions are sought is through state regulations and legislative processes and such processes are, therefore, instructive in this regard. According to Stenius, 'Nordic political culture sees laws and legislation as wearing the mantle of sacrosanct venerability' (Stenius 1997: 170), adding (albeit with specific reference to Denmark and Sweden) that there is a kind of pragmatism which recognizes that society can be changed and that 'these changes should be brought about in particular by instituting new laws' (Stenius 1997: 171). This postulate might be overstating the issue, as new laws or changes in existing legislation are also a consequence of people's actions. The example I analyse in Chapter 5 (regarding homosexual parenthood and surrogacy) is a case in point. Nevertheless, Stenius's claim that the universality principle presupposes an active and interventionist state is brought out in legislative processes – also those that concern procreative practices.[37] Thus, conversely, legislation also plays an important part in instituting the state, forwarding an image of consensus, and in the process glossing over the contentious nature of some issues.

There is another point to be made with regard to the Norwegian state – and that is the relationship to the Norwegian Lutheran Church, which is also a state Church, implying that the state governs the Church and the Church of Norway has a privileged position (Thorkildsen 1997: 154). This privileged position is iterated in the Norwegian Constitution. Article 2 states: All inhabitants of the Realm shall have the right to free exercise of religion. The Evangelical-Lutheran Religion shall remain the official religion of the State. Article 4 states: The King shall at all times profess the Evangelical-Lutheran religion, and uphold and protect the same.[38] Although there is a process underway to revise Article 2 of the Constitution, Norway still has a state Church, and that Church plays – and has played – a significant role in Norway, also politically. For example, until 2008 – after having lost a case in the European Court of Human Rights (see Borchgrevink 2009) – Norway had a Christian object clause with regard to public education. This implied that schools and kindergartens were to help parents raise their children in accordance with the basic values of Christianity (see Leirvik 2004). Norway is nevertheless a secular state, ensuring freedom of religion and confession.[39]

In his discussion of religious identity and Nordic identity, Thorkildsen argues that Lutheranism has been one of several important factors that have contributed to the construction of the Nordic welfare state. Moreover, he posits that the churches became inscribed in processes of nationalism (which he describes as the new civil religion) where 'Christianity and the Nordic national churches became part of a historical heritage' (Thorkildsen 1997: 157). This is an interesting observation as it points to the role of Christian values (and not necessarily belief) as intrinsic to a national – or secular – culture, a transformation that also has repercussions for politics. Witoszek supports this contention, stating that, 'Swedish, Norwegian and Danish nationalisms require to be re-read through the prism of Christian values' (Witoszek 1997: 81).[40] This point can also be brought to bear on the legislation pertaining to biotechnologies (see especially Chapter 5), although as we shall see it is not always easy to differentiate between what are specific Christian values and what are the values that emanate from a cultural heritage based on Christianity. It is therefore interesting to note that 'our cultural heritage' or 'Western cultural heritage' (*vestlig kulturarv*) is often referred to in terms of a specific Christian heritage, without actually specifying what this is, but rather implying that we (Norwegians that is) already know. However, what is important to note is that the Church and those somehow affiliated with it (albeit indirectly) have a voice with regard to the legislation of biotechnology. They are 'heard' *inter alia* through its various bishops, through the Church Council, the Church Meeting; through the Faculty of Theology (University of Oslo) and the Norwegian School of Theology (Menighetsfakultetet); the Christian Doctors Association and, of course, through the Christian Democratic Party. Notable also is that many of the early bioethicists in Norway were theologians. Thus, there is a significant political influence. However, and this is important: the Church does not speak with one voice. In fact, the Church is split on issues regarding the ordination of female priests and bishops; on the right of homosexuals to marry; on abortion; on the use and application of reproductive technologies; and on research on embryos.

Thorkildsen claims that the two central pillars of the state – full employment and social security – correspond closely with central ideas in Lutheranism: daily work as the fulfilment of God's vocation and 'a priesthood of all believers' that 'promoted a culture of equality, where obvious wealth and large social differences were not acceptable because fundamentally all individuals are equal and have the same worth' (Thorkildsen 1997: 159). Stenius likewise stresses the importance of work – or that everybody works – as the force that holds society together. However, the significance of work lies not only in the fact that it is inclusive (Stenius 1997: 167) but also that charity is considered humiliating, as is the means testing of the poor (Trägårdh 1997: 254). In contemporary Norway, this implies *inter alia* that disabled persons should be able to participate fully in society and not be dependent on the good will of others.

The above points to certain significant ideological aspects underpinning the organization of the state that bears relevance for grasping the importance of legislative processes. These have to do with universality, trust in government, and the role of the Church. They do not explicitly concern equality, but they do underwrite an 'ethos of equity' that in some ways is attributed to the state and also emanates from it, so much so that state and society are often conflated.[41] Included in this ethos is the emphasis placed on individual autonomy.[42] The notion of autonomy is also closely linked to the vernacular term *folk*, which takes on a special meaning in Nordic languages. *Folk* refers to ordinary people and is often associated with being authentic, inclusive and more trustworthy than the *elite* (or experts). Hence the term *folk flest*, literally 'most people', is a common vernacular idiom that connotes a silent majority that effectively resists paternalistic forms of governance (Lien and Melhuus 2009: 201). The political legitimacy of *folk flest* may seem to contradict an extensive trust in government. Yet, in Norway, this is not always the case. Partly this may have to do with an ideology that politicians are, if not 'of the people', at least no better than *folk flest*. Partly this may be explained by politicians' self-presentations: not only are they as *folk flest* but they also extol the virtues of common sense (as against, for example, expert knowledge: see Chapters 3 and 5). Whatever the case, there is overall an acceptance of state intervention, and state interference into private and personal matters is the rule rather than the exception.[43] However, the justification of such intervention will necessarily evoke some notion of equality – be it in terms of *rettferd* or *likhet*.

The Norwegian notion of equality is used by many Norwegians to describe Norwegian society. It informs official national self-presentation and permeates socio-cultural processes in a variety of ways. The significance of the notion of equality in Norway is not whether Norwegians are more equal than people elsewhere, but is, as Vike, Lidén and Lien (2001) suggest, rather that this notion is rarely challenged. Moreover, as Vike, Lidén and Lien demonstrate, the notion of equality is sustained and strengthened through various forms of institutionalization, such as health and social policies, educational policies, labour market and industrial policies, and consumer and cultural policies.

However, they stress that the apparent cultural, social and economic equality that pervades Norwegian society may as much be a result of nation building and redistributive policies as it is of any inherent or traditional cultural homogeneity (Vike, Lidén and Lien 2001: 18–19). In fact, the idea of cultural homogeneity is effectively undermined by the presence of the Sami and more recently immigrants from countries whose cultural and religious backgrounds differ markedly from those of ethnic Norwegians. However, in both cases the notion of 'equality as sameness' has played an important role in the state policies (assimilation and integration) that have been pursued.

Vike, Lidén and Lien (2001) echo some of the same concerns as Sørensen and Stråth (1997a). Moreover, although they recognize the value of equality as an empirical phenomenon in contemporary Norwegian society, they suggest that its meaning is changing. In their own words: 'today equality is as much about agreement on just rules ... as it is about equality as a result' (Vike, Lidén and Lien 2001: 26). Obviously, equality is as much about producing difference as it is about producing sameness, and it is an empirical question whether this notion will retain its status as a gate-keeping concept in studies of Norwegian society. My study of biopolitics based on the legislative process of reproductive technologies can be read against this grain. Although the notion of equality does not frame my analytical perspective, it is apparent that some meaning of this notion also informs different aspects of the legislation and the people involved.

The politics of reproduction link many social fields and, as I hope to show, they are a potent site for political contestation and resistance, not only in the present but also in the past (see also Franklin and Ragoné 1998a). Assisted reproductive technologies are evocative in ways that other subjects of legislation would not be. They broach ontological, epistemological and ethical issues. The stakes are high as the elements involved not only go to the core of the liberal state and its dilemma (individual freedom versus state control) but also strike at the very cords of life itself and what it is that creates meaningful relationships. These have to do with such things as the status of the embryo, the question of rights, the significance attributed biogenetic bonds and a biological principle, as well as issues of knowledge production and competence building and questions related to the public health system – in short, the very reproduction of society. Writing just over a decade ago, Strathern says: 'In the two decades during which reproductive technologies have become a public phenomenon, both [citizen's relationship with the state and the ethics of intervention] have been arenas for what one can only call cultural exploration' (Strathern 1999: 65).[44] Another decade has passed, and it seems that this arena has yet to be tapped for what it may reveal about contemporary societies. The following chapters constitute one more contribution to this cultural exploration.

Chapter 2

CHILDREN OF ONE'S OWN

> Among axioms central to Western culture has been the conviction that the family is rooted in the very nature of things. Definitions of 'family' and of 'nature' vary. In the Western world, however, consensus has existed that the family was as natural as the turning of the sun, as immutable, subject as strictly to intrinsic law, as indispensable to the conduct of life.
> —Janet Dolgin, *Defining the Family*

A First Encounter

I attended my first meeting and seminar of the Association for the Involuntary Childless (Forening for ufrivillig barnløse, FUB) towards the end of 1998.[1] This was my first encounter with 'the field'. In the course of that day, my attention was particularly drawn towards two people, a man and a woman. They were both very active during the seminar, and especially so during the general assembly. They were extremely knowledgeable about the various issues being discussed, and more specifically concerned about the medical and economic aspects of infertility treatment. One issue that they raised was the question of tax deduction for expenses incurred in conjunction with infertility treatment (they had paid large sums) and how to cope with the problem that different case workers would reach different conclusions. Besides the fact that they were especially articulate, what attracted my attention was that they were both wearing a black armband.

I was told that they were a couple and that they had both been active members of FUB for many years, holding office and generally involved in the running of the association. They had been in infertility treatment for ten years without results and they had now finally decided to stop. They had 'tried everything' (as the saying goes) and had even been to an infertility clinic in Belgium. I approached them and asked about the armband. The answer was poignant. They wore the black armband, the man told me, to make visible the loss of a child they had never had. They were, he claimed, in a state of mourning and wanted that fact to be known. It was their way of making a public statement about their grief.

This small but highly symbolic act condenses the many feelings and, not least, experiences of what being involuntarily childless means to many of those concerned. It evokes the pain – and invisibility – surrounding this condition and the processes involved in coming to terms with the fact that the condition will not be overcome. More forcefully perhaps, it testifies to how very difficult it is to find adequate ways of sharing the loss of something that never was; of 'burying an unborn child', as the woman said, indicating the emotional turmoil, suffering and frustrations entailed. The black armband symbolized an end, the culmination of a process which had irrevocably marked their lives; a process, I was to learn, driven by one desire: to have a child, and one obsession: to do what it takes.

Sometime during the morning session of the seminar, a woman walked in with a baby in her arms, leaving the bright new baby carriage conspicuously in the doorway. Heads turned and someone rose to greet her. Others looked away. During coffee break several women gathered in one corner. They were clearly upset. 'She shouldn't have come', I overheard one woman saying. 'No. This is just too much', added another, while a third complained 'She should have known better – to show up here with her baby!' Throughout the day I heard similar comments, all indicating that it was not only unwise of her to come, but also inconsiderate – almost indecent. The young mother was actually accused of showing off her baby, 'flaunting it', as one woman said.

From the snippets of conversation I picked up, I learned that she had been through many IVF cycles and had finally succeeded in getting pregnant and carrying her baby to term. She had also been an active member of FUB and knew many of the people present. It was therefore surprising – to me, and maybe also to her – that rather than welcome her and include her and her newborn baby in the event she was practically ostracized. Rather than representing hope and a visible proof of possible success, she and her baby's presence were perceived as a provocation, unleashing negative reactions and emotions. For many of those present it was just not possible to share – or even acknowledge – her feelings of joy and pride. She was welcome to participate if she wanted to, but she should have left her baby at home.

This was not only my first encounter with the people and problems that were to engage me for many years, but also one which in many ways laid the basis for the subsequent directions of my fieldwork and research. I was (literally) introduced to this world through members of the Association for the Involuntary Childless. This is a non-governmental organization, founded in 1982, that (according to its statutes) serves the interests of people who struggle to have children. The occasion was their yearly business meeting, which was held in conjunction with a seminar. The seminar lasted two days (a Saturday and Sunday) and the venue was a location in downtown Oslo. The seminar covered a range of topics thought to engage the involuntary childless. The opening talk was given by a social anthropologist. Her theme was kinship and reproduction in light of the new reproductive technologies.[2] The next speaker was a psychologist who addressed the psychological aspects of childlessness. He was followed by a medical doctor, one of the pioneers of infertility and IVF treatments in Norway. Thereafter there was a presentation on hormone stimulation by a representative from Serono, one of the main pharmaceutical companies providing drugs for infertility treatments. The next speaker was a woman who spoke about acupuncture and its merits for the treatment of involuntary childlessness. Finally there was a debate about whether or not involuntary childlessness is an illness (with the implication this classification has for public health services), with the participation of politicians, patients, doctors, the government health administration and members of FUB. The general assembly closed the first day. By that time there were only about twenty people left.

Having an 'Own Child'

The two anecdotes – the black armband and the rejection of the young mother – are emblematic of the kinds of emotions that involuntary childlessness entails. I was to come across them again and again during my conversations with the various women (especially the women) and men who self-identified as involuntarily childless. These emotions – and the ambivalences they express – not only form the backdrop against which their ideas and attitudes to assisted reproduction must be understood. They also fuel their acts and their relationships with others. To a certain extent this emotional undertow underpins their views of which procreative practices are acceptable and desirable. To be confronted with the fact that you are unable to have children when you have assumed that 'they would come naturally' represents a turning point in the lives of those concerned. How you choose or are able to cope with this fact will necessarily vary. Some are able to assume their infertility and choose to live without children. Others are unable to accept that verdict and move towards an uncertain process of trying to become parents, to become a family. The women and men I talked to in the course of the following years all belonged to the latter category. They all wanted a child they could call their own, an 'own child' (*eget barn*) as they would say.

However, what an 'own child' is and what it means is not given a priori. The meanings ascribed to this notion will vary depending on a series of circumstances, the most important of which has to do with time. In fact the temporal dimension is intrinsic to any understanding of what involuntary childlessness entails for those concerned, where the 'biological clock' is but one (albeit important) aspect of this process. Eventually, it is the possibility of having a child at all that gives content to the idea of 'own'. It is the yearning to be a family, to have children 'just like everyone else', which drives the involuntary childless and underpins the various processes that engage them. For some it may take years in order to reach the longed-for status of becoming parents and being a family; years marked not only by uncertainty, ambivalence, hopes and disappointments, in short, continual strain, but also by constant decision making. Thus time intercedes and leaves its marks in terms of events, emotions and attitudes. Time – or rather the passing of time – upsets the content of this desire and contributes towards the transformation of certain ideas of what kinship is and what it should be. Facts and values are stirred; the relation between what is and what ought to be is challenged as alternative forms of relatedness are considered. These considerations are complex, ranging from issues of health (what are the risks of sustained hormone treatments?) to economic concerns (of the available options, which can we afford?) to the more troubling questions of what is best, right, for them and for the child.

Their point of departure is to conceive and give birth to a baby that is biogenetically related to both parents. That is considered the natural way and constitutes the basic, taken for granted, understanding of 'own' children. If a couple is not able to conceive, one of the various forms of assisted conception using ova and sperm from the intending couple is the first possible option.[3] To all intents and purposes, the resulting child is 'the same' as any other child, in the sense that its relation to its parents in biological terms conforms to traditional understandings of family connectedness premised on blood relations. Its natural creation only needed some assistance, and this form of reproductive assistance is now culturally accepted. However, if that option is not possible (for medical reasons) the issue then is, on the one hand, whether or not to use donor gametes and, on the other, whether or not to adopt. It is at this juncture that what constitutes an 'own child' starts to move, shifting from its biogenetic foundation to a more socio-cultural one. This shift may be barely perceptible, yet it is nevertheless fundamental.

The way forward towards having an 'own child' will necessarily imply a disruption with the assumed natural (f)acts of kinship. The possible alternative forms of relatedness require a rethinking of what 'own' means. They also require a reconsideration of other fundamental relationships: the relationship between biology and love; between biogenetic connectedness and emotional connectedness; between natural facts and social facts. Under normal circumstances (if we can still use the term 'normal'), these are conflated. They come together, as it were. It is presumed that you will love your own (biological) child. It is presumed that blood is thicker than water. However, once the biological connection is out of the equation, either partially or totally, what remains is the emotional bond, and the desire to have a child (cf. Strathern 1995: 355).

Many of the women and men I talked to were concerned about different forms of bonding – about family and what it means to be a family, and about kinship and what it means to be kin. These two were not necessarily understood as being the same, although one (kinship) is seen as an extension of the other (family) (Leira 1996). Many explicitly – and even spontaneously – stated that it is being with and caring for the child that counts; it is the daily care and continual involvement with the child that makes the relation.[4] This was in their terms 'natural' and presented as a matter of fact. However, biogenetic connectedness is also valued (naturally), given an emotional content, especially in terms of belonging and kinship reckoning more widely – that is, in relation to potential grandparents, aunts, uncles and so on. One of the most visible articulations of the meanings of biogenetic connectedness is expressed in terms of kin likeness – that is, in terms of physical resemblance. For some, kinship is manifest first and foremost through physical resemblance. To look alike is to be kin (å være i slekt). A biogenetic connection ensures (or acts as a guarantor for) a certain genealogical embeddedness which carries its own significance. This form of relatedness implies a certain knowledge that has to do with origin: knowing where (that is, from whom) the child issues and by implication to whom it belongs. As we shall see later (Chapters 3 and 4) such knowledge has come to play a major role in another context: that of making the law.

It is hard for me to say whether these contradictory notions are held more generally (they probably are) or are a product of the very situation of being involuntarily childless. What is evident is that the situation of having to choose how to have a baby makes such notions explicit, producing a self-conscious discourse on kinship and relatedness. This is a salient characteristic of all those involved in these processes. Once it is known that having a child the natural way will not be possible, this knowledge produces its own effects – it prompts a process of reflection that, in addition to incredulity, shame and even blame, broaches questions tied to the meaning of life, being a couple, and what having a child – any child – implies. This process is undergirded by the FUB through their various activities – such as annual meetings, occasional public meetings on specific themes and the various talk groups that they (at that time at least) helped arrange. However, the point I wish to make here is that these various notions of nature and nurture not only exist, they co-exist – within the same person, couple or even situation. People struggle to make sense of their own understandings and feelings, recognizing fully that they are not necessarily consistent. The different meanings ascribed to social and biological connectedness are mobilized according to context and the shifting explanations take on the semblance of a figure/ground motif: the one serves as backdrop for the other. These two notions are mutually constituted, mutually implicated. They reflect the core of the nature/nurture configuration as it is articulated within kinship and assisted reproduction in Norway.

An Issue of Sameness

In what follows, I explore some of the meanings attached to the idea of an 'own child'. After first giving some general considerations, I present excerpts from conversations with five different couples in order to illustrate the variations in attitudes to and their preferences for the available alternatives of assisted conception. Their arguments will be rendered more or less verbatim, as this better illustrates their rationalization, in terms of how they approach the matter and in terms of the values disclosed. These narratives are personal and context bound and are, at this level, unique. They are individual tales about a shared theme, indicating a certain complexity in their particulars. However, it is possible to discern a common reasoning underpinning some of these stories. This reasoning merits attention in that it reflects (at another level) some central values within contemporary Norwegian society. These have to do with the notion of *likhet* – understood both as 'sameness' and as being equal.

Although my overall thrust is not primarily to demonstrate how such notions operate, my ethnography also speaks to this issue, in this chapter as well as in ensuing ones. Thus, I bring forth some other illustrations of the workings of this ideological tool in contemporary Norway. To anticipate the argument: there is an idea of an 'imagined sameness' (to use Gullestad's term) that is articulated in people's attitudes to what an 'own child' implies in terms of relatedness. In the case of the involuntary childless, it is the idea that the prospective parents each somehow have the same relation to their child that is notable. The understanding of what 'same relation' means will, in turn, guide their decisions as to what option to follow when it comes to forms of assisted procreation – be it through the help of reproductive technologies (with or without donor gametes) or transnational adoption. Moreover, the negotiations surrounding what possible options to pursue also reflect on gender relations, equality and how parenting is understood and managed as a mutual project, or not (see also Lappegård 2007).

What is striking in the Norwegian case is that many involuntarily childless people consider adoption a better alternative than the use of donor gametes. Keeping in mind that adoption in Norway implies transnational adoption – as very few Norwegian born children are put up for adoption (see Howell 2006) – this is noteworthy. In so far as these children come from China, Columbia or Korea (and not Russia or Romania) the child will usually look very different from its parents. There are two important, albeit different, reasons for a preference to adopt.[5] If a couple decides to adopt a child (and granted that they are accepted as potential adopting parents) they know that they will eventually have a child – or, to use the literal Norwegian expression, 'be given a child' (*å få et barn*). Adoption implies certainty, and hence a fulfilment of the desire to have a child. This can be an important aspect as assisted conception cannot guarantee a 'take-home baby'. Moreover, in terms of biogenetic relatedness, an adopted child will – by its very lack of this bond – have the same relation to both its adoptive mother and father. In other words, adoption ensures that the child is equally theirs – in the sense that it does not belong more (or less) to one than the other. It is this idea of being the 'same' that entails a reckoning of equality in terms of a presumed balance in the relationship between each parent and the child which is prominent. This reckoning has a double message: it says something about the relationship between parents and children as well as the relationship between the parents. Moreover, it tells us something about the importance of biogenetic connectedness by the significance attributed to its absence.

Before turning to the conversations, I want to make one more point. The process of overcoming infertility – of becoming parents – is in some senses paradoxical: on the one hand, the women involved (and it is first and foremost the women, as it is they who embody this experience) described the process as if it was driven by its own momentum; they are drawn in, but not by their own volition. On the other hand, the same women (and some of the men) are extremely articulate about what they are doing and why. They are reflective and the producers of a self-conscious kinship. They have learned that you cannot take for granted even the most natural of processes and have thought about what this implies. The excerpts below capture this mood.

An 'Own Child': A First Approximation

Most of the women and men I talked to were in the process of undergoing infertility treatment. Some had already had several cycles of treatment (either in Norway or abroad); some were just starting; some had been successful and some not, at least not yet.[6] I also had the chance to speak to women who had 'been through it', as they said, but were now past that stage. Some of these women had had a child; others realized that their chances of becoming pregnant were indeed very slim and were struggling to come to terms with that situation; and yet others were considering adoption.[7] The latter were either couples who had been told – or advised – by their doctors to stop treatment or couples who for various reasons preferred adoption to assisted conception. However, most of the couples I spoke to who were considering adoption had first tried in vitro fertilization (IVF). There were also couples who now formed part of what is termed reconstituted families – that is, couples where either one or both had been through a divorce and had children from former relationships. Hence, those I spoke to were reflecting on the issues of 'own' and 'belonging' from different positions.

The information I have gleaned about the meanings of 'own child' derives from various situations and contexts. Sometimes (as in the case of the interviews) my questions have been direct, and in part hypothetical, as when I asked them to consider the various options (whether or not they are legal in Norway or even available in Europe, as for example surrogacy).[8] In other cases, the information is more indirect, as in the talk groups I participated in where the conversations would be much more unstructured and covered several themes in the course of an evening[9] or during a meeting (arranged by FUB).

The involuntary childless themselves are well aware of all the available options (be they legal in Norway or not), and most of those I have spoken to are extremely knowledgeable about the various treatments, where they can be obtained, their costs and so on. In fact, the sharing of information is one of the major activities of FUB, and talk groups are established for those who feel the need to share their particular experiences, disappointments, frustrations – and there are many. Retrospectively, it seems that the notion of 'own child' is one of those taken-for-granted notions that only become explicit and susceptible to scrutiny because of the very disruption of the natural process of procreation that not being able to have children entails.

To Try Everything; To Tell or Not To Tell

Once it has been established that there is a problem with conceiving naturally (usually after one year of systematic trying) and the couple has decided to do something about it, they enter into a process that may take years before it is successfully or unsuccessfully terminated. Having a child becomes a major project in their lives and for some it becomes an obsession, so much so that they withdraw from family and friends who have children. Indeed, many women related that they cannot bear to be in the presence of children, especially babies, and pregnant women. Their relationships with their surroundings become strained. Nearly all that I have talked to said that there is little understanding or sympathy to be had from friends and family. In fact, most of them claimed that a person who has not experienced what it is to be involuntarily childless – and in particular people who have children – cannot understand their situation and are basically not entitled to an opinion. Interestingly, this attitude was reflected by several of the politicians I talked to who felt that it was very difficult to make decisions which would affect the involuntary childless seeing as they have themselves been blessed with children. Many of the involuntary childless said they felt like social outcasts and find Christmas and 17 May (the Norwegian national holiday and, like Christmas, a children's day) particularly painful. They would flee the city on such occasions.

There is one point on which all the involuntary childless concur: to not have children, to not be a family is very difficult in Norwegian society today. Family values are important in Norwegian society, even as the family as an institution is undergoing changes. Children represent access to significant social – and socializing – arenas from which those without children are effectively excluded. Thus, for many, to be voluntarily childless, it seems, is practically a no-choice situation. You are expected to try everything (that is, infertility treatment) and you are expected to at least have thought through and made a decision about adoption. In other words, there is a socio-cultural expectation on people to have an informed position. At least the involuntary childless themselves feel these pressures and feel they have to act accordingly. As one woman said: 'It is good to feel that we have tried everything. Instead of sitting there twenty years from now ... At least we have tried everything'. Or, as another voiced it rhetorically: 'When is enough enough? How do I justify an adoption?' And another commented, 'The technological development can make you regret what you did not do'. Whether it is their natal family, in-laws or friends that are the most significant reference points varies, but that the external surroundings only serve to exacerbate inner feelings of failure, even shame, and incompetence is evident. Moreover, feelings of incompetence and failure are often coupled with a rhetoric of unfairness and injustice. Why me? What have I done to deserve this?

However, being involuntarily childless is not necessarily a fact that couples feel they can share with others. Hence, one of the first decisions that has to be made is whether or not to be open about their condition. Many choose to remain silent, and carry the burden on their own, apparently shrugging off jokes or remarks such as 'It's about time you have children', or 'Are you only interested in your career?' or 'You are lucky not to have children and can do what you want'. Others choose to be open in the hope that they might then receive empathy (but resisting anything that might resemble pity) and also avoid the standard moral condemnation: that they are egotists, only thinking about their careers or wanting to spend money on expensive cars and holidays rather than on having children.

Perhaps precisely because infertility is a sensitive subject that many couples find difficult to discuss openly, the problems that emerge between a couple seldom surface. In the course of my conversations it became evident that men and women have different attitudes to the lengths they would go to in order to have a child. This was especially the case if the male partner had children from another relationship. As Julia said: 'He can manage without children. He has two. But I want a child, our child'. And she had to work very hard to make this child a mutual project and not just hers. This not only entailed having 'sex by the clock', as she put it, but also (later) persuading her partner to come to the clinic (with her) and produce the necessary sperm. Marianne was in a similar position: she felt her relationship to Odd's ex-wife would be easier if she, Marianne, had a child with Odd (as his ex-wife has a child). Yet Odd could not understand why this was so important to her and refused to be drawn into the process. However, these were exceptional cases. More common was the regret (on the part of the women) that their men offered little sympathy; that they (the men) did not understand; that they did not want to talk. Nevertheless, many couples said that they were in this together and the men tried as best they could to be involved – even if they did not necessarily grasp the extent of the emotional costs and, not least, the women's emotional instability (in particular when taking hormone treatments).

Even though most of the people I talked to phrased their childlessness as a mutual problem (and would often present it as such) the decision to remain silent seems to be adopted especially in cases where the 'problem' was the man's – that is, if it was a question of poor sperm quality. According to some women I talked to, exposing the male partner as infertile is more problematic than exposing the female partner. This is a bit hard to understand considering the expectations placed on women to bear children and become mothers, and an underlying cultural assumption of female identity tied to pregnancy and birth. One could assume that for a woman it might ease her situation to know (and let others know) 'the problem' is not hers. This is in fact the case in two of the examples that I give. Even so, there are perhaps some unexamined assumptions about the link between sperm quality and masculinity that are not often voiced. Male infertility can be resolved by sperm donation, whether anonymous or not, and is both easier and less painful than other forms of assisted conception involving IVF. However, sperm donation is not seen as an easy option. In part this may have to do with the secrecy surrounding anonymity – which may also imply that those who were planning to use anonymous sperm would not relay this fact to me.[10] It may also have to do with the introduction of a third party into the relationship (which could also apply to egg donation, but does not in the same way). However, the third-party argument is not just about an 'unknown father' or the idea that there are two fathers, it is also about skewing the potential relationship that the parents have with the child. Whatever the case, there was evidently a sensitivity on the part of some women to 'cover' their men, along with the feelings of failure, incompetence and shame which run through many of these reproductive histories.

What Makes a Child Your Own?

How, then, do the involuntary childless – when confronted with the various possibilities that assisted conception offers – imagine different forms of relatedness? And how do these imaginings influence their actions? My considerations are prompted by their stated desire for a 'child they can call their own' and by my question as to how you make a child 'your own' or 'what is it that makes a child your child?' In the following conversations with five different couples (in two of the cases the husband was not present) I pursue these and related questions. These particular examples have been picked because they illustrate the many dimensions of being involuntarily childless and the complex negotiations that take place. They also indicate some interesting differences between them, reflecting the values that, in each situation, are given most weight. Finally, taken together they also articulate some significant insights about ideas of relatedness and how these are articulated in contemporary Norway. Included in these is the conjugal relation (whether those interviewed are formally married or not).

Anne and Jon

At the time of our conversation Anne is 37. She is an engineer, and has been married to Jon, also an engineer, for nearly eight years. Jon was at home, but did not want to take part in the interview, something Anne regretted. They have been trying to have a baby for almost nine years and when I meet Anne they have started an adoption process, having passed the first hurdle, which is to be accepted as potential adopting parents.[11] Yet Anne is wistful. They had been to the adoption office and the caseworker there had told her that she must overcome her grief (of not being able to bear a child) before she goes ahead with the adoption. Anne comments that it is so hard to let it go, and admits that it is very difficult for her to come to terms with the fact that they will not be able to have their own child – that is, a child that comes from them, as she says. Anne has a daughter from an earlier marriage, yet she is determined to have a child with Jon. They both want this very much and had, at the time of the interview, tried everything – including clinics at home and abroad, homeopathy, acupuncture, magnetic mattresses, foot therapy and a telepathic healer. In their case, there is a problem with Jon's sperm count.

Anne told me that she was in fact relieved that it was her husband's sperm quality that was poor. She was afraid that her in-laws would think that because she already had a child she was refusing to have one with Jon; that, in effect, she was being an egotist. Therefore, atypically, she wanted them to know not only that they were trying through assisted conception but also that Jon was the one who had the difficulties. For her, it was paramount to be accepted by her in-laws – and having his child, as she put it, would give her a place in that family. She also said that it would have been much harder if she were the one who was causing the problem, and she would have had a very bad conscience and feel guilty (towards him).

Knowing this background to Anne and Jon's case, I asked Anne whether they had considered donor sperm:

ANNE: Yes, we have. But it is top secret. We could never tell anyone, if we do that. Not to the family. We could never, never tell. Not to anybody. Not even to the child. So it is like living a lie ... Here [in Norway] the child does not have the chance to find its father, so if the child should find out [that it has been conceived by anonymous sperm] it would never be able to meet its biological father.[12] If I were to find out that I had a biological father somewhere, I would do everything I could to meet that man.
M.M.: Why is that, do you think?
ANNE: People are different ... I have very strong feelings about finding your roots and where you come from ... I want to know everything, meet the flesh and blood of my origins.
M.M.: Did you discuss this between the two of you?
ANNE: He is positive and I am negative. I find it difficult ethically. He doesn't quite understand that. If I had been him I would have thought the other way round: [this way] I would be having a child with someone else. So I tell him: I do not want to have children with anyone else. It would be much more fair to adopt. That is the way I think.

Anne does not want to have a child that is not also biologically Jon's – that is not 'theirs'. If she cannot have the real thing, she feels it is more fair (*realt*), as she puts it, to adopt. This has to do with her ethical views both with regard to the child (and its need to know its origin) and with regard to her husband (she does not want to have a child by another man). Nevertheless, she is also wary of adoption, for two reasons. On the one hand, she is afraid that the child will encounter problems in Norwegian society because it will look different. 'They won't know he is Norwegian', as she says. She fears that it will be mistaken for an immigrant child. 'We know we will love it, but we cannot know if life will be good to it'. On the other hand, and more forcefully, she is very much concerned about kin likeness. She wants a child who will look like her husband (as this is her gateway into his family) and she feels that if they arrive with an adopted child her in-laws will feel deep inside that the child is foreign 'as it does not look like any of us', as she says. For her it is easier to accept surrogacy – with his sperm and her eggs – than it is to accept the use of donor gametes. In her case, it is the biogenetic substances that count as the fundamental bond. 'Surrogacy is quite OK', she says, 'because it comes from Jon and me ... but it has lived with another woman, has grown there, but it is really ours'.

When I ask Anne whether they have considered a foster child, she answers:

> Actually, I have thought quite a lot about that. We have the resources and the time to have a child every third week or so. But we want our own child, a child that only we, if I can put it that way, own. Not really own, but that only has us. We do not just want to be with children, we want a child in order to have something together. A child that only has us ... and that we have together. Now I am not really one who thinks that a child is ours, because I look at each child as very independent ... as an individual.

For Anne an 'own child' means one that only has them – the parents. It is the exclusive relation that is significant – and that it belongs to both of them. Ideally, this child should have both their genes, it should 'come from' both of them. The fact that Anne already has a child and is a mother influences her thinking. She is very much concerned about being accepted by Jon's family and it seems that having a child with (if not by) Jon is paramount in that regard. Adoption is an option that provides an own child – one, which in her words, will only have them (as parents). She is silent about the adopted child's biological parents (and about that child's right to know its origins). In her view, adoption is a good alternative because it is 'more fair' – it does not skew the balance between her and Jon. Yet as her project is to become more integrated in his family, she is wary of pursuing this option wholeheartedly, as she is not sure the adopted child will be fully accepted (neither by Jon's family nor by society more generally). Anne's oscillation between the different alternatives expresses the ambivalences and concerns that many involuntarily childless people have, and not least how difficult it is to give a particular action priority.

Maria and Alex

Maria echoes some of the same sentiments voiced by Anne. Maria is 30 and her husband, Alex, is 32. She is re-educating herself as a nurse, and he works as a salesman. When I met her the first time, she was 'in treatment', as she said.

When I interviewed them in their home some months later, she was seventeen-weeks pregnant. After she and Alex discovered that they were unable to conceive, Maria was first told that she had a problem with her fallopian tubes. She felt this as humiliating – and a defeat. To her, it was an assault on her womanhood, not being able to get pregnant. So she was relieved – in a way – when they found out that Alex had an extremely poor sperm count (due to an infection that he had had for a long time). She had always wanted a child very much and had never imagined that there would be a problem. In fact, when Alex says 'If it had been the other way round; that we had first found out that it was my problem, she would have left me', Maria admits that that would probably have been the case: 'I know it sounds grotesque', she says, 'but I think I would have'. Later she comments: 'I don't think that we could be happy, just the two of us … there is a lot one can do without children, but …' She leaves the sentence trailing. She continues: 'It is me who has done everything. If I had not kept at it as I did, we would not have had any test tube [prøverør]'. Alex adds: 'No, we would not be where we are today'. 'No, we wouldn't', Maria concurs, adding, 'And it has been very tough'.

Earlier in the interview, Alex had expressed his own frustrations:

> This situation has been worst for Maria. She has carried the biggest sorrow. I have not been able to show my sorrow as she has, and in a way she has carried the sorrow for both of us … It has been terrible to see her pain [he is here referring to the injections, hormone treatment and removal of eggs]. So I feel that in a way it is my fault, because it is me that has such poor sperm quality. I only want the best for her. So I could think I should leave her so that she can be happy, find someone else that she can … but it just doesn't work that way, does it?

Alex was not positive about sperm donation; he would prefer to adopt. He has tried different things to improve his sperm quality, such as acupuncture, alternative medicine, extra vitamins and so on. However, it was not made clear exactly what treatment they had accepted, and whether or not they might have made use of donor sperm but were not willing to say so.

Maria's desire to become pregnant and become a mother was part and parcel of her life project, and this was very apparent in her comments about the ethical aspects of assisted reproduction. Answering my question about the limits to the application of reproductive technologies and who should be able to decide, Maria says: 'Well, really … the ethical. I couldn't really care less. I am a Christian and all that, but when it comes to this, I am willing to go through whatever it takes. As long as one can have a child I don't give a shit about ethical concerns. Maybe it is just because I choose not to sit down and think about it. I will do anything'. She then likens her situation to one who has heart trouble and says: 'If I had a poor heart I would of course accept a pig's heart, right? It all has to do with the situation one is in'.

On the question of using donor gametes, Maria says (and in contrast to Anne): 'I think that if the eggs are donor eggs I would think it was my child, even if the eggs belong to another woman. If we adopt a child ... it would be as much your child as an own child'. Maria juxtaposes egg donation with adoption. She is not so concerned about the biogenetic bond; rather, she is concerned that a child has one mother and one father, and that it is the parents that mould the child. It is a matter of care (*omsorg*), as she says. By way of illustration, she uses the following example:

> If he had a child from an earlier relationship which I got to know when it was five, let's say, this is not my child. That child already has a mother and father – a mother that it lives with. It could not be my child. I could become a step-mother ... but to adopt a child, that would be my child. The parents of that child have chosen not to keep the child and in that case it is in fact us who become its parents ... [A] child needs a mother and father. Whether they are biological or not, does not make any difference.

In her explanation Maria points to the difference in relationships that emerge in a reconstituted family. She cannot be mother to a child that has a mother. Adoption is different, because the mother (or parents) has relinquished all claims to the child. The child will be theirs as the relationships it mobilizes are, in her understanding, unequivocal. In this sense she concurs with Anne and the importance of creating exclusive relationships between the parents and the child, a child that has only them and that they have together. Maria uses the same kind of reasoning regarding foster children; there is also here a certain ambiguity as to the types of relationship that are established. In many ways, Maria perceives adoption as unproblematic; at least she does not voice any major concerns. Maria wants to have more than one child and she – as well as Alex – say they will consider adopting a second child. 'But we must first see ourselves through this'. They do not think that they will treat an adopted child any differently than one that is biologically theirs. As we can see, Maria's position is not consistent. On the hand, she states that she would have been willing to leave her husband because of his infertility problems; yet on the other hand, she (and Alex) are very positive regarding adoption. Such inconsistency is not uncommon, revealing the ambiguities that being involuntarily childless gives rise to, and the shifting grounds that decisions are made on.

Susanne and Dagfinn

Susanne and Dagfinn, also a couple in their thirties, had been living together for several years before they started thinking about having children. They both have good jobs and enjoy travelling. They have long felt the expectation and pressure of friends and family to have children. According to Susanne, their friends think that she and Dagfinn have the wrong priorities, waiting to have children until they are forty. Dagfinn comes from a rural background where it is common (among his friends) to have children in their early twenties. 'My old friends, they do not know what childlessness is', he says. They stopped using contraception while refurbishing a house they had bought. That process took almost two years and at that time a friend suggested that perhaps they should have a check-up just to make sure everything was in order. They are now very glad they did. It turned out that Susanne had had some cell changes and that Dagfinn had a low sperm count. When I meet them, they are 'in the system', as she puts it, and being treated. As they have not been totally open about the fact that they have had problems getting pregnant, they still find their relations to others strained. 'People are so inconsiderate – they think we do not want children', Dagfinn says. Susanne feels that it might be easier to tell them that they do want children but aren't able to: 'It might be easier to tell it like it is. That would stop them from making snide remarks (*slenge med leppa*)'.

They are both positive about sperm donation and have considered the possibility, as long as the donor is anonymous. They do not want two fathers. On this they both concur. Dagfinn says: 'I am positive too [to anonymous sperm donation]. It is something special for a woman to be able to carry and give birth to a child. Another alternative is adoption of course. And if things should turn out that way ... but it is closer to be able to give birth to a child'. Susanne then adds, without my prompting, 'Well, this method [egg donation] is not permitted in Norway, but egg donation would actually be the best because then the woman would be participating at the same time as the husband is. Both are more participatory than with semen donation'. Susanne finds it strange that the egg is more precious (a term she uses) than sperm: 'I mean egg and sperm have the same amount of genes. I don't understand this politics'.

They have also considered adoption, and Dagfinn says that this would be the next step. In fact, he says they have considered adoption all along, but wanted to try IVF first (at this time they are being treated by ICSI). 'We have friends who have adopted, and they say, emotionally, that to fetch a child [in a foreign country] is as big as giving birth', Dagfinn says. Susanne is more reluctant. She is not as positive towards adoption as her husband:

> All I have said is that I do not know how I will react if I have three failed attempts.[13] You say that then we just adopt. But perhaps, I have said, perhaps, if we feel we still have a hope, we should try one more time. I am not quite yet finished with this process – to give birth to my own child, to put it that way.

Dagfinn says he will be the father whatever they do, 'whether it is IVF, sperm donation, I will be the same father for that child'.

Being a mother, being a father, being a parent – these matters lie at the heart of our conversation:

> M.M.: So what is it that makes you the father?
> DAGFINN: The fact that we are together, that we are doing this together, whether she gives birth or we fetch a baby; that we raise it.
> SUSANNE: It is about being a mother and father. It is about being able to contribute something. You will mould (*forme*) the child, even if it is an independent individual.

DAGFINN: And that is irrespective of whether it is assisted conception, adoption or whatever.
SUSANNE: Our greatest fear if we adopt is that we might get a child that has a 'scar on its soul'. That is what we are really afraid of. That we will not be able to give it the security it needs because it has experienced so much deprivation (*svik og omsorgssvikt*).

They are not so concerned about kin likeness as they are about giving the child a good home and being able to be good parents. Like Anne, however, they also wonder whether an adopted child would be accepted into the kin group. They agree that nothing negative will be said, but they nevertheless wonder if there will, inadvertently, be a differential treatment by the prospective grandparents of an adopted child and their other grandchildren (which are not adopted).

Dagfinn accepts that it is possible to live a full life without children, yet he says: 'If this process does not work, we will adopt. But the thought that we might end up being alone and still have a good life is not foreign'. Susanne: 'Yes, I am sure we would, but with a child we would have someone to care for, that is our own'.

In the case of Susanne and Dagfinn it is her desire to give birth to a baby which is the overriding concern and which guides her decisions. She would rather use donor gametes than adopt, as long as she can give birth. Surrogacy, 'this American thing' as they call it, is not an alternative. Both state very clearly that it is their 'being together' about the child (and I understand them to mean being in agreement about the way the child is to be had) and raising it, that it is a mutual project (those are my words), which is fundamental, and which makes the child theirs. Neither of them stresses the biogenetic bond in particular, but both have thoughts about adoption. Although they are positive towards sperm donation, they insist on anonymity. This is to avoid there being two fathers, as Dagfinn says. Susanne is likewise, if not more, positive to egg donation, as this allows the spouses a more equal participation in the creation of the child than does sperm donation. This notion of equal participation is one that many others also voiced, and tells us as much about the relation between parents and child as is it does about relations between spouses. One is contingent on the other. Perhaps I should add that about a year after this interview took place I received a card from the two of them: Susanne had given birth to twins.

Judith and Paul

Judith was one of the first people I interviewed and echoes some of these same thoughts. When I ask her, towards the end of the interview, at which her husband is not present, what it takes to make a child that belongs to both, she answers:

I think you have to have the same interest in the child ... the same attitude ... Both desire the child a lot, work for the child, both wish to make an effort for that child

... It does not have to be the same effort ... in the same area all the time, but that you work at it so that it does not end up being: this is your child, you fix it. If one is going to adopt both must want it. That is why I did not want to adopt when he did not want to. It would have been a 'this is your child' situation and I would not have been able to face that. So I would hold that a precondition is that one has the same attitude – that we are together on this child. It has to be that way no matter what one chooses. Also with donor semen ... I think that with donor semen the mutuality [togetherness] of the couple is more difficult to obtain than with egg donation. The man is the one who always looks on. It is the woman who gives birth, right? If it is the partner who has given the semen, than he is part of it. If he has not, then he is just the woman's husband, in a way. If it is egg donation it is the man's semen and the woman carries the child. Throughout pregnancy and birth there are very strong bonds that develop even if the eggs are not genetically your own.

Judith is 53 and her husband, Paul, 54. They have a daughter; she is (at the time of the interview) a teenager. Paul has two children from a former marriage. Judith is also divorced, and it was during her first marriage she realized that she might have trouble getting pregnant. That was also the cause of the divorce. Judith then met Paul, and they have been together since. Judith is a scientist and, as she said, thinks like a scientist: she was convinced that once the problem had been identified, it would be solved. Her fallopian tubes were blocked and it was just a matter of opening them up. Then she would get pregnant (the normal way). However, as it turned out, it did not work out that way, and she started her first IVF treatment as early as 1983/84, when the methods were much less developed than what they are today. During all this time, she avoided people with children if she could. She said: 'I felt like an amputated woman, I felt I did not function like a woman. I was not a real woman. I only looked that way'. She continues:

> Around 1984/85, I thought that I would never get pregnant – I no longer had a perspective on my life. I felt that with children everything would seem so much more relevant. Everything that goes on in society becomes more relevant; everything that goes on in the world becomes more relevant, right? So I had decided that this is over with. I then received a letter from the hospital telling me that I was too old and that they could not give me any more tries. I was then 38.

A few months later Judith was pregnant, without assistance.

When I ask Judith about how she and Paul handled the situation from the beginning, she said: 'The problem was with me and I knew that then. If we had not had our child we might not still be together. It would have been difficult for me ... he has his own children that he loves very much, and I think it would have been difficult. I would have preferred to live on my own'. Even retrospectively Judith admits that she would have had problems relating to him and his children if they did not have a child of their own. And she adds 'These are his children; they have their own mother. I did not want to push myself upon them. I have grown up with divorced parents and am the first child of my father's first marriage. I have never called any of his later wives "mother"'.

I ask Judith: Did you ever consider adoption? 'Yes', she replies, 'for me that could have been a solution, but not for Paul. He said: "I have children, what do I want with a foreign child?" So that was no option'. Judith reflects on the difference between adopted children and 'own born' children and whether she would treat them differently. She says:

> This is really hard to imagine. Almost impossible. I think one should treat all children the same. I also think that it should be possible to have your own children and adopted children. I think that it is not right that just the involuntary childless can adopt. To me it seems as if adoption in Norway is used as a treatment for childlessness and not to give a child without parents a home. I think that is wrong. When it comes to adoption there should be more focus on helping the child.

When I ask her more about motherhood, the mothering instinct (*morsinnstinktet*) and adoption, she answers: 'I think maybe that if you have not carried a child the mother would have the same [kind of] relation to the child as the father ... I read once in a psychology book that a mother's love is unconditional, but that a father's love is conditional. It is possible that a woman that adopts a child has a more conditional love'. She is, as already mentioned, positive about egg donation, and says that, with the development of ICSI, sperm donation will probably disappear. However, she insists that in the case of sperm donation the donor should be anonymous. That is best for all those concerned.

Judith is looking back at a period in her life which has been painful and full of emotional turmoil. She recalls the situations and her reactions as if they were yesterday. There is no doubt that her experience as one of the involuntary childless has left a mark. Yet she is – and was – an active woman with a good job and many interests. She finally managed to get pregnant 'on her own' but has nevertheless continued to be an active and involved supporter of FUB. Her reflections underscore a preoccupation that many have. She is basically concerned about how a man and a woman (as potential parents) will have the same relation to a child, so that the child will not belong more to one parent than the other. (This becomes all the more clear through her reflections on her relation to Paul's children.) In her opinion, egg donation permits this form of relatedness, where carrying and nurturing the child on the one hand, and his sperm on the other, balances the equation, so to speak. This is an idea that Susanne also expresses. Judith (along with many of the others I interviewed) stresses the importance of the conjugal relationship: that the way you choose to have a child must not undermine the relationship between the husband and wife. The option chosen must not produce an 'it is your child, you take care of it' situation. And she voices sentiments tied to reconstituted families that many other involuntarily childless people echo: that if one of the spouses (or partners) already has a child or children it actually makes the situation more difficult; in addition to belonging to the spouse, these children also 'belong' to someone else, and therefore not only upset the conjugal relation but may also 'remind' the infertile partner of his or her 'failure'.

Ola and Kari

This same consideration of the conjugal relation was also voiced by Ola and Kari. They stress that the problem is theirs – together. They will be open about their struggle to have a child, but they will not say who is the cause. They are very conscious about not placing the blame on either of them. Yet it is Ola's poor sperm quality that is the problem.

Ola and Kari are a young Christian couple; both are under thirty. She is a school teacher and he is pursuing an academic career. They have been married for five years. When I meet them for the interview, they are in the middle of treatment. His sperm is of very poor quality and they have been offered ICSI as a possible solution to overcome his infertility. At this time (the late 1990s) ICSI is still a novel procedure in Norway (and still only provisionally permitted), and they are both very uncertain of the outcome. They have therefore had many conversations about other possible alternatives, but adoption appears to be their most realistic option. Moreover, due to their religious convictions, they will not permit the destruction of embryos; hence they will only allow the medical personnel to fertilize the eggs that are to be reinserted. They are well aware that this lessens their chances, but they are adamant: an embryo is life and therefore cannot be destroyed. They have told their parents about their situation and Ola has also informed his brother, thinking that his brother should know in case he also might have the same problem.

They stopped using contraception a year after they married. Ola says: 'We had decided to wait a year. We did that, but then of course nothing happened'. When I ask Kari what her reaction was when they were told that his sperm count was very low, she answers:

> KARI: We stood together, all the time. No matter whose fault it was, it was something that we were in together (*som vi var to om*). So that one of us could not come and say, 'but it is your fault that we cannot have children'. This burden is heavy enough as it is ... it is a heavy burden whether it is me or him who is the cause.
>
> OLA: We talked about it before we went for the examination, that this is a mutual problem. For me it was important, that if it was Kari that was the reason [for the failure to conceive] that I would not blame her for anything. So I had really worked hard on that question: if it were she that was the cause [for our infertility], and not me. So it was painful ...
>
> M.M: How?
>
> OLA: I had all sorts of thoughts in my head; the next fifty years of our lives, we will not be able to produce one single child. That is what we had been told. It was the rest-of-life consequences for me. So it was quite a shock. That there is something fundamentally wrong with me; that surprised me. I am used to standing up for who I am, do things even if they are unusual ... But this, it was something that was very painful.
>
> KARI: In a way, I was glad that it was not me, but not because I wanted to blame him. It is not that. What actually was most important was that we found out what it was.

In their situation they are most likely to go for adoption as they are not very optimistic about the results of IVF and ICSI. They do not wish to use donor sperm. Kari, who has been positive about adoption all along, says that if she is going to give birth the child has to be theirs – that is, genetically theirs. She finds it problematic that the child will be related not only to her but also to an unknown father (she uses this term). Ola says when prompted:

> Yes this has been a foreign thought to me. I need to have a rational explanation for what I do and I have asked myself why [I cannot accept sperm donation]. I realize that I am inconsistent, because I am all for adoption, and in that case neither of us is genetically connected. It is perhaps better that we are equal (*likeverdig*). Either both or neither. I am afraid in a way that if it is her genes and not mine ... I won't be as much of a parent as she is. I think that is a reason, even if I know it is inconsistent because with adoption I am convinced that I will be able to see it as my child. I don't know why it is that way, but I just feel it.

One of the issues that invariably arose in my conversations with the involuntary childless was the question of money, the economics of it all. In Norway, a couple is granted three cycles of IVF treatment using the public health care system. If a couple decides that they want more than this they have to go private, and that costs money. Adoption is also costly, and in addition adds another two years to the period of waiting. Thus, one consideration that was often mentioned was the fact that if you go for adoption you are at least certain of the outcome: you will have a child and your money will be well spent. Kari expresses this sentiment: 'We have pamphlets [from private clinics], but I feel it is too much money right out of the window if it doesn't work. I would rather spend that money on adoption. Then you are at least sure that you will have a child'. And she adds: 'To be adopted means that you are a very much desired child. That often becomes much clearer with adopted children than with own born (*eget født*) ones. An own-born child may even be unwanted'.

Ola and Kari have no qualms about the use of reproductive technologies or in principle the use of either donor sperm or donor eggs. As Ola says: 'Politically I find it very inconsistent to allow sperm donation and not egg donation. Either you are against both or open to both. I do not agree to this idea of the wrong egg in the wrong uterus .... The way I see it, this is blatant gender discrimination'. And later in the interview he adds: 'This thing about nature and nurture. That has also been part of the political argument for not permitting egg donation. I feel that they are muddling two things. One thing is what one will accept politically and ethically, but another thing is what the woman herself is willing to do. If a woman wants to carry a child that is not hers genetically, it is no business of the authorities'. He argues forcefully that childlessness is a private matter and that people should be able to choose what suits them best.

As with the others, Ola and Kari also stress the importance of being together on whatever course they choose. They are on the whole positive about the range of assisted conception techniques, though constrained by their religious convictions when it comes to the destruction of embryos. With regard to the significance of genetic connectedness, they express different views. For Kari the question is whether she is to give birth: in that case, she wants the child to be genetically theirs with no third-party intervention; otherwise, she is for adoption. Ola, however, recognizes his inconsistency – about being for adoption but against sperm donation – and his explanation is that it is best that they (as a couple) are equal, either both or neither. This is in my understanding what ultimately underpins the parental role – and frames the notion of an 'own child'.

Drawing Together

So what do these snapshots of people's private thoughts about how they envision their relation to a child tell us? First of all, and perhaps obviously, that there are different attitudes to these technologies and that the meanings conveyed are not necessarily consistent. Reproductive technologies provide options that couples are willing to use, but the various options are considered in conjunction with what a child – or having a child – means to them, and what kind of relationship they feel it is important to have to the child. Central to these considerations is the significance of being pregnant: what it means to the woman – and man – that they be able to bear children. Moreover, the conjugal relation – and how it is perceived – represents a fundamental dimension in their reflections about what options to pursue. In this regard the following is noteworthy. Underlying the different attitudes to the various options and combinations available, there is one idea about relatedness that stands out: the importance of creating a child that in some way is perceived as equally theirs – equally shared. However, it is not necessarily the contribution of similar or equal substances that make for this equality. Rather, in their idea of imagined sameness, each parent must somehow occupy a position with respect to the child that can be understood as similar or the same in some crucial aspects. Otherwise they risk skewing the parent–child relationship in favour of one or the other, and thereby also, in their understandings, undermining the conjugal relation. An 'own child' embraces a sense of belonging and participation that implies an exclusive relation that no others can be privy to. Thus, egg donation is in some cases viewed more positively than sperm donation. This sense of an exclusive 'own' and belonging is also expressed with regard to the children of a spouse or partner of another marriage – they 'belong' (at least partially) elsewhere.

A biological model with its notion of a fifty–fifty contribution of genetic material is a main frame of reference; this is the 'natural' point of departure. As Howell states when commenting on her research on transnational adoption in Norway: 'biology remains the model by which most Norwegians approach kinned relatedness' (Howell 2006: 38). This model concurs with – or can be seen as an expression of – a bilateral kinship system. So much so, that some of those interviewed explicitly state that they do not understand the differential treatment of sperm and eggs in the legislation. The positions offered can be understood as variations of this fundamental understanding. Yet, the option to adopt, although making both parents equally unrelated to the child, in practice denies the significance of biogenetic connectedness. Hence, it is evident that biogenetic substances do play a significant role in inscribing certain forms of relatedness; so much so that the absence of shared substances may be as forceful an argument as the presence of either own egg or sperm. Moreover, the various combinations that people envision as appropriate equivalences are interesting, as is the general acceptance of using either donor sperm or donor eggs – or even surrogacy – if that were possible. Couples will accord different aspects of nature and nurture different values – such as a donor egg plus the husband's sperm but in the wife's body – but the overall aim is to achieve a balance that gives the semblance of equality and unity within the family.

In imagining their future baby, their future family, each half of involuntarily childless couples takes into account not only their own relation to the child but also their relation to each other. In that regard, adoption (which negates the biogenetic bonds between adopting parents and their adopted child) is a good option because each parent will have the same relation to the child, neither being biogenetically connected. An adopted child works so as to ensure a balance between the parents and the unity of the family. Moreover, and in contrary to assisted conception, the outcome of adoption is certain. These are significant factors. Yet, adoption involves other considerations. The adopted child will most likely look different from its parents, thus not only raising the question of kin likeness but also making evident (in the sense of making public) problems of conception. Moreover, there is also the fear of discrimination – within the kin network and in society more generally.

For the majority of those I interviewed, a move to adopt comes in the wake of failed IVF treatment and is a result of an uncertain and often painful process. Thus, their decision to adopt must be seen in this light. Nevertheless, as Howell demonstrates, adoptive parents successfully kin their children through their efforts 'to make their adopted child into *their* child and a relative to their own relatives' (Howell 2006: 64). Thus adoption is an option that perhaps most visibly illustrates the efforts that over time go into creating an 'own child', through processes of incorporation and transformation (see Howell 2006). The ambivalences – for example, of Susanne versus Anne regarding the use of egg donation – and the sometime inconsistencies – such as those of Maria, who was willing to leave her husband if she could not give birth while at the same time being willing to adopt – revealed in these interviews with regard to the different options is an indication of the complexities involved and how difficult it is to come to terms with being involuntarily childless – and not least to finding a solution that it is possible to accept, for both, in order to have a child they can call their own.

At a general level, my topic concerns the tie between possible forms of procreation and filiation, and how certain ideas – in this case among the involuntary childless – of significant relatedness informs the nature of this tie. In this vein, reconstituted families throw a particular light on these issues, articulating some of the various ideas of 'own' and 'belonging'. I have, in this chapter, examined how parents reason with regard to decisions about the forms of relatedness they wish to have to their future offspring. This reasoning includes perceptions that have to do with meanings of ova and sperm and how these meanings feed into other significant relations – most notably that of marriage or cohabitation and parenthood – but also towards parents and parents-in-law. Certain perceptions of kinship relations underpin their considerations. However, the images that these couples project are in many ways different from – even inconsistent with – those embedded in the law. This is in itself an important datum, and one that will be pursued in the following chapter. Moreover, in contrast to the imaginations that ground the law, which can be said to embrace a public space, these images of parent–child relationships are particular and individual. Nevertheless, just as with the law, albeit in a very different way, they too reflect underlying values in Norwegian society. In addition to the value of being a family and thereby included in meaningful sociality, this also involves an idea of equality – articulated as a kind of imagined sameness.

In the next chapter I shift perspective. The focus is on legislative processes and how various reproductive technologies have (or have not) been incorporated into Norwegian law. It traces a legislative process that commenced in the early 1950s and continues through to 2007. Thus we move from those subjected to the law to the very subject of the law.

Chapter 3

BETTER SAFE THAN SORRY: LEGISLATING ASSISTED CONCEPTION

> Ethnographic literature ... shows that new reproductive technologies are often viewed as morally threatening, disrupting not only notions of personhood, parenthood, and family formation, but also the moral fabric of society as a whole. Interestingly these moral disruptions are not culturally invariant. If there is one finding that has emerged most clearly from this recent comparative ethnography in First World settings, it is that new reproductive technologies can evoke varying moral responses from one society to the next.
> —Marcia Inhorn, *Local Babies, Global Science*

> This is a reproductive cycle in a novel sense. A child is technologically conceived who embodies the need for a law; a law is brought into being because of this child's birth; the law will bring into being other children who will embody it.
> —Sarah Franklin, 'Making Representations'

Uncomfortable Relations

In the introduction to her book *Kinship, Law and the Unexpected*, Strathern notes that in some places law and biotechnology work together whereas law and kinship often do not, in that notions of the embodied and distributed person sit uncomfortably with the legal subject (Strathern 2005: 10). An interesting point to pursue, ethnographically, is the form that these working relations take. The Norwegian law has, by consistently prohibiting egg donation, confronted the problem of the distributed person by partially negating the very possibility of its creation (in Norway). I say partially, in that sperm donation has been practised and tacitly allowed since the early 1950s (and even before) and explicitly since 1987.[1] Nevertheless, it appears that sperm donation does not create the same unease with regard to fatherhood as egg donation does with regard to notions of motherhood. This point will be pursued in the next chapter.

As we have seen in the previous chapter, there is an uncomfortable relation between law and certain practices of kinship (through assisted procreation) in that those legal subjects directly affected by biotechnology laws are not necessarily in agreement with what the law prescribes, and are, moreover, willing to circumvent the law if need be. In their efforts to legislate, lawmakers, although acutely aware of the plight of the involuntary childless, do not only – or even necessarily – have the involuntary childless in mind. They see their job as making laws that are – at least to some extent – congruent with their visions of and obligations to society. At its most abstract level, this implies legislation that embraces society and restricts developments that may in some way challenge fundamental values, such as respect for human dignity, human rights and non-discrimination. Central to the debates – and the various positions taken – are the views held of science and technological developments: who has the right and obligation to intervene in these, and on what grounds. These views will obviously vary, not only as individual personal convictions but also across party lines. As legislation is a political process, different interests are also involved. Such differences are evidenced in the debates in Parliament about biotechnology (and the processes leading up to these) as well as in other parts of the public sector. Reproductive technologies strike at the core of basic life processes, and also involve a potential restructuring of kinship relations, as these have been understood. Hence, the debates are framed as ethical, and ethics is seen as a premise for the very same debates. The application of a precautionary principle has to be understood in this light.

One tendency underwriting the relation between law and kinship (as viewed through the regulation of reproductive technologies), I argue, is that of maintaining certainty – prohibiting egg donation removes any doubt about who the 'real' mother is.[2] This is articulated through a belief in the truth and effectiveness of biological relatedness (Salazar 2009). Thus, the uncomfortable relation that Strathern envisions takes a different form in Norway. So also with regard to the relationship between law and biotechnology. In the Norwegian case there has been a public will to govern the applications of biotechnology and thereby discipline its practitioners by allowing only certain forms of practice. The law works to restrict the potential uses of biotechnology; hence the relationship between law and biotechnology, although explicit, is an uneasy one.

With the advent of biotechnology and reproductive technologies, conception – that is, the very preconditions for bringing life into the world – has become a social event with new meanings attributed to it. Whereas the moment of birth is a visible one, and an event that is perhaps universally an occasion for public recognition of some kind, moments of conception have been invisible – both physically (not being visually available) and socio-culturally. However, developments in reproductive technologies have radically changed both these conditions. In vitro fertilization (IVF) has rendered conception visible, and is now considered a routine practice in reproductive medicine. These procreative practices have in turn been the focus of public attention in a variety of ways.

Thus, reproductive technologies have literally visualized conception and placed it at the forefront of public consciousness. Both legislative bodies and the media have contributed to this awareness. One effect of such attention has been to familiarize, even naturalize, these acts of conception. Another has been to moralize these biomedical practices, in the sense that they are actively drawn out of a scientific realm and placed firmly within a moral universe. Thus, such procreative practices are conducive to and activate debates about the good, right and natural. Finally, and perhaps most significantly, we are, as Strathern observes, 'not surprised if intimate medical matters concerning third parties are debated in public' (Strathern 2005: 17).

As in many other personal matters that in some way involve sexual relations – illicit sex, doubts about paternity, illegitimate children – so too with assisted procreative practices: society finds a way to speak back. This may take different forms. However, whereas issues involving the transgression of sexual mores are often commented on and discussed in the privacy of homes or within intimate circles, only to surface publicly through gossip or slander,[3] questions involving reproductive technologies speak forcefully and legitimately in public.[4] This publicity is multifaceted. Sensational headlines and films capture in different ways the moral dilemmas these technologies pose. Legislative acts and the rise to prominence of bioethics are other ways that such concerns have been channelled. Whatever the expression, with the advent of assisted conception and reproductive technologies, matters of conception have entered the public realm with a particular force, in Norway as well as in many countries all over the world (see Ginsburg and Rapp 1995b; Inhorn and van Balen 2002).

Reproductive technologies reconfigure the public/private divide precisely because they draw attention to issues of conception and procreation in a way that has made them obviously part of what Norwegians term *offentligheten*, or public sphere. Both as a social phenomenon and as individual dilemmas, they are placed squarely in the public domain.[5] Reproductive technologies have set an agenda that mobilizes a whole series of agents – not least the state and the Church, but also scientists, medical practitioners and the involuntary childless themselves – each of whom has a particular stake in the issues involved. Because of their potential to intervene in life-creating processes, these technologies have spurred imaginations not only about the very constitution of personhood and the distinction between human life and human being but also about the kind of society that an uncritical incorporation of these technologies might imply. In Norway, such imaginations have prompted the application of a precautionary principle in legislative matters concerned with biotechnology and reproductive technologies, and produced the term 'sorting society' (*sorteringssamfunnet*), a term which carries persuasive power and is especially evocative, a point that will be developed in Chapter 5.

The Legislative Process: Acts and Revisions

This chapter focuses on one significant aspect of this public arena, namely, the legislation of assisted reproduction and the concomitant legislative processes that have fed into the formulation of the Norwegian Biotechnology Act (and its revisions). Starting with the first attempts to regulate artificial insemination by donor (AID) in the early 1950s and culminating in the revision of 2007, this chapter covers a time span of over fifty years. During this time the first law regulating assisted reproduction was passed in 1987; it was revised in 1994 (being incorporated into a biotechnology act), and revised again in 2003 and 2007. These legislative acts are events that both pin down and sum up a socio-cultural process. They are also events that foster debates. In what follows I am as much concerned with the processes leading up to and surrounding these events as I am with the event itself. In addition to presenting the main contents of the acts, I will in particular focus on legislators' attitudes to the legislation.

The fact that the laws regulating assisted conception have been revised several times (to date, in 1987, 1994, 2003 and 2007) says something about the working relationship between law and biotechnology. It indicates the political and contested nature of these technologies and their associated practices. A shift in government has implied a shift in legislation. It also tells us that with regard to this particular subject matter legislation is seen as a meaningful way of dealing with developments within science and technology. One may well ask whether these revisions are a characteristic of the legislation or of the technologies involved. Legislators I talked to would say that it has to do with inherent traits of the technologies and the science that produces them: there is an innate drive among scientists to continually renew and refine technologies, making possible procedures which were unthinkable just a few years earlier. As new knowledge displaces the old, it is the task of legislators, I was told, to (minimally) keep pace, preferably being ahead of such technological developments (Melhuus 2005). Whatever the justifications, biotechnology and law work together and are mutually implicated. And, as will become apparent, one aspect of this relation is tied to legislators' attitudes to knowledge.

Legislative acts and not least the debates that surround them cast a focused light over a turbulent social field and reveal something about the values under dispute. Moreover, the very framing of the debates gives a certain direction not only to the way the issues are perceived but also to the way they ought to be perceived. They direct the public's attention towards particular alternatives (to the exclusion of other possible alternatives). In so doing they establish a norm for how fundamental kinship relations should be understood. In other words, they set an agenda.[6] What makes the law regulating assisted conception especially interesting is the fact that this was a new area of law. The Artificial Procreation Act of 1987 was the first law to be passed which addressed these concerns – indeed, it was also one of the first in the world.[7] This legislation was prompted by the birth of the first Norwegian-born IVF baby in 1984. This event mobilized both politicians and the public at large. Among legislators, there was a general consensus that something had to be done. The politicians cum legislators were very much engaged in the process, and all those I talked to stressed the excitement (and even privilege) of being involved in the creation of a new law (something that will be discussed below).

However, as early as 1953 – that is, years before this law was debated, formulated and eventually passed through Parliament – there was an attempt to regulate the practice of artificial insemination by donor (AID), as it was then called. At the time, donor anonymity was the rule. Strictly speaking AID is not a reproductive technology; it is a practice of assisted conception. It is also the practice that has been the most controversial and where significant shifts of opinion have occurred. Many of the arguments put forth then have been repeated since, with some success, especially those related to the use of anonymous donor sperm. Other arguments – for example, those tied to the fundamental centrality of the institution of marriage – have lost their significance precisely because the context has changed and meanings and values have shifted with the times. I turn now to this effort to legislate AID with a specific emphasis on the discourse that framed the debates. Although this attempt at legislation was not successful, it is nevertheless important as it foreshadows the discussion to come as well casting a historical light on later processes.

Artificial Insemination by Donor: For and Against

In 1950 the Norwegian government appointed a committee to review the questions relating to AID. The committee consisted of five people: a state bureaucrat, a priest (later bishop), a medical doctor, a psychologist and a housewife.[8] This was part of a Nordic effort to coordinate legislation regarding AID. It was known that artificial insemination was being practised (Løvset 1951),[9] and the committee was asked to evaluate the need and conditions for artificial insemination as well as the legal status of the child. The committee was asked to cooperate with similar committees in other Nordic countries.[10] The committee submitted its report in 1953. The majority suggested that legislation be passed permitting artificial insemination with anonymous donor sperm. The priest and the housewife expressed their disagreement, each submitting their own dissent. However, no legislation was passed.[11]

The 1953 report opens with a discussion of the legal grounds for AID, drawing specific attention to the following: the legal status of the child (focusing on the issue of legitimacy); whether artificial insemination by donor could be considered a sexual offence; whether AID could be considered a case of adultery; whether the practice of AID would fall under the law of quackery (*kvakksalverloven*); whether the registration of a child born by AID would amount to forgery; and the question of consent. These legal issues are more or less resolved in the report. The committee states that insemination can be considered a sexual offence only if it is done without the woman's consent. Insemination does not fall under the law of quackery; neither can the registration of a child born by insemination by donor be considered forgery (*dokumentfalsk*) as long as the husband has agreed and the rule of *pater est* holds.[12] The committee argues that the woman (who is being inseminated) cannot be charged with breach of the marriage contract as adultery implies sexual relations with another man. Further, the committee states that a man who gives sperm is not guilty of breach of the marriage contract even if he does so without his wife's consent, although such an act may be considered a violation of her person (*krenkelse av hustruen*).[13] Throughout its deliberations, the committee takes for granted the institution of marriage, and moreover assumes the principle of *pater est*. In fact, this principle was the basis upon which the other arguments rested. These arguments were first and foremost moral, and the most controversial issue was that of secrecy implied by donor anonymity. It is this controversy that has reverberated down to the present, culminating with the rescinding of donor anonymity in 2003.

The debates for and against AID in the 1950s shifted between biological and psychological arguments, between nature and nurture and the moral connotations these evoke with respect to what is natural, good and right.[14] Arguments on both sides centred on the institution of marriage, the home and love, on the one hand, and women's natural desire to have children and the significance of infertility for male identity on the other. The act of lying and the introduction of a third party in marriage are themes constantly reiterated. Not surprisingly, those persons who are for regulation of AID give weight to nurture over nature. Regarding the father–child relationship, their main argument is that what is important for a child's psychological development is not its biological origins but its home environment, which, more specifically, should be a harmonious home. A good home, moreover, was seen as consonant with a good marriage. The thrust of the argument is that relatedness is not necessarily born of biological bonds. Adoption and step-children (with reference to the role of the step-father) were used to support the committee's statements, as in the following: 'If the conditions in the home are good and the husband actually assumes the position of father in the rearing of the child, there is a strong chance that the child in its being and character will take after the husband/father in such a way that he [the husband] will gradually feel that he has a true (*virkelig*) descendant'.[15] At the same time, the majority of the committee underscore the significance of the donor child's biological relatedness to its mother as an important element that will contribute towards strengthening the bond between the spouses.

Those who were against AID turned the arguments around. They point to the fact that homes are not harmonious and that divorce is increasing. Furthermore, they insist, that a child biologically linked to its mother and not its father could create tensions between the spouses rather than foster unity. Most significantly, they claim that introducing a third party into a marriage is equivalent to undermining that very institution. To quote the minority, Bishop Smidt: 'Our Christian culture rests on the home. Monogamous marriage is the very foundation of our culture. With insemination a third party is introduced into a relationship between those two, that should be one, and between parents and children. Here one has driven a wedge into the very life principle of the home ... It may be that this is one of these deathblows (*grunnskudd*) that our culture cannot tolerate'.[16]

Although the arguments for AID tend to be couched in terms of nurture, there are nevertheless also biological (or 'nature') arguments in support of AID. The most important of these are related to the woman's possibility to become a 'real mother'. The desire (*trangen*) to become a mother belongs naturally to marriage and contributes towards realizing marriage's full potential (*mening*).[17] Again adoption is used as a point of reference and comparison, but now as a contrast to that of true motherhood. It is assumed that motherhood (understood as giving birth) will give the woman a value (understood as life quality) both physiologically and psychologically that cannot be realized in the same way through adoption. At the same time, it is argued that insemination by donor allows the husband to overcome his feelings of inferiority (attributed to him as a result of his infertility) in that his wife will bear him a child that in the eyes of society is his (biological child). A tangential argument, which is nevertheless significant in light of later developments, is that the number of children put up for adoption was steadily decreasing. (Transnational adoption was not then on the agenda.) This has to do with ideological changes at the time: On the one hand, it was seen as immoral that a mother should have to give up her child;[18] on the other hand, the shame attached to being an unwed mother – and having an illegitimate child – was waning. Nevertheless, according to Leira, the postwar years (1945 to 1960) was 'the prime time of the conventional nuclear family in Norway', and '[traditional] family values embracing the home-based mother/housewife flourished' (Leira 1992: 103). In a certain sense, then, the housewife (*husmoren*, lit. 'housemother') embodied a morality of marriage, motherhood and the home. This morality was in turn reflected in views of family and kinship.

The institution of marriage is central also for the majority of the committee.[19] The arguments for legalizing AID rests on an idea that insemination contributes to realizing 'the essence of marriage, which otherwise would not be consummated (*fullbyrdes*)'.[20] They argue further that legislating AID will familiarize people with this practice, thereby contributing towards its demystification, and not least to killing the myths circulating at the time: that AID is a form of state-controlled human breeding, against the order of nature, a sin against the sacredness of marriage, and so on.

By the early 1950s, artificial insemination by donor was already a public issue. While the committee was working on its report, a heated debate was going on. This debate had in part been provoked by the recent publication of a book by the bio-psychologist Rønne-Pettersen which likened assisted conception to magic (Rønne-Pettersen 1951).[21] At about the same time, a Norwegian gynaecologist published the results from his research about patient's attitudes to artificial insemination (Løvset 1951). Løvset's aim was to put forward information that could contribute towards the future practice of AID.[22] Although the conclusions are not entirely clear, Løvset's position is: childless couples should be helped, and AID is one way of helping them. A well-known Norwegian writer and essayist, Axel Sandemose, also wanted his say in this debate, publishing an essay entitled 'Conceived in Lies' on the subject (Sandemose 1952).

Both Sandemose and Rønne-Pettersen are adamantly against AID. Whereas the latter considers artificial insemination to be the equivalent of rape, in that it is a sexual act devoid of emotional content, Sandemose insists that marriage has a biological purpose (*hensikt*) and to change this purpose is tantamount to undermining the very nature of man. For both of these men the main concern is the relation between spouses and fundamentally the institution of marriage, as well as the relation between sperm donor and sperm receiver. And the recipient in their perception is not the woman being inseminated but her husband, who is infertile. In their view, adultery is to be preferred to AID. Not only is a sperm donor an irresponsible person (and therefore not suited to procreate), but also, according to Sandemose, the act of donating sperm is a homosexual act (Sandemose 1952: 26). A sperm donor gives his sperm anonymously to other anonymous men. Thus, for Sandemose, a sperm donor enters an unnatural relation between men, expressing a specific form of male bonding. This argument is similar to ones that feminists (at a much later date) have voiced, although with a different twist: anonymous sperm donation hides male infertility; thus men protect other men, in order to uphold masculinity and safeguard the patriarchal nuclear family. As we shall see, the later debates about the practice of anonymity take on yet different arguments, although the question about lying, which was very prevalent in the 1950s, remains important for some.

The arguments of Sandemose and Rønne-Pettersen in particular are extreme.[23] Yet, I include them because they echo the main concerns of the overall debate and hint at the predominant climate framing the arguments. They make very evident some essentialist perceptions of male and female, not least regarding the fundamental purpose of marriage. For Sandemose it was the act of lying and the secrecy that anonymous sperm donation entails that was most problematic. Although the question of what it means to be an 'insemination child' (as he puts it) was raised, this was not his main concern.

The question of secrecy was also a principal concern of the committee. Again parallels are drawn to the adopted child and the possibility of finding out that it is adopted. Precisely because the practice of informing a child or not of its adoption is inconsistent, they argue, AID is preferable as it is most likely that the secret will be kept.[24] They are also concerned about the psychological effects that the knowledge that father is not 'father' might have on the child. Even though the problem is considered serious (if it should arise), attention is directed at whether such an eventuality would occur. Two scenarios are put forth: in the case of conflict between the spouses and in the case where insemination itself might cause marital problems. And this strikes at the heart of the matter.

Those who are against legalizing AID consider the introduction of a third party into the marriage a 'violation of the very sacredness of the home'; the uncertainty about the child's origins is in itself sufficient to destabilize the marriage. The 'insemination child', to use their terms, although very much desired, will cause marital discord. It is not possible to erase the husband's feelings of inferiority through a mock or pseudo solution (*skinnløsning*). He will be reminded daily of his incompetence; 'he has by his side a wife whose motherhood is real, while his fatherhood/paternity is false'.[25] In spite of recognizing a woman's desire to have a child, it is immoral, they argue, to let 'curative considerations' determine the birth of a child. They stress that the precondition of a successful marriage does not rest principally on having children. On the contrary, the value of marriage rests on the personal relation between the spouses. For them, adoption is considered a good alternative if the desire for children is strong – not only because adoption ensures that both parents will have the same relation to the child (an argument we have seen echoed by the involuntary childless today), but also because the adopted child already exists and the adopting parents can give it better possibilities than it otherwise would have.[26] Insemination is considered an intervention in a natural process and to permit such intervention can have unknown consequences. Again, adoption is put forward as a better alternative because 'the adopted child's creation is natural, just as adoption is a natural and familiar practice, and it does not involve an intervention in any biological mechanism'.[27]

Perhaps the most important argument against legislating the practice of AID is that AID 'conceals the truth and proscribes secrecy about the mutual relations amongst a series of people ... and ... departs from the principle of biological fatherhood'.[28] This argument is linked to the postulate that legislation will generate a general doubt about origin and that 'the insemination child will lack true knowledge of its origin and therefore also of itself'.[29] Although this identity issue was not predominant at the time, it is precisely this argument that gains prominence in the 1990s: a child's right to know its biological origin is viewed as a precondition for knowing who he or she is.

Bringing the Past to the Present

I have rendered the debate of the 1950s in some detail in order to get across some of the ideas that prevailed at the time. This is important for several reasons. First and foremost, it is significant that there was no consensus on the issue of AID. As we shall see, issues tied to assisted conception and associated technologies have been consistently contested over the years. As with IVF, and later pre-implantation diagnosis and research on embryos, AID triggered imaginations. It is the fear of what these practices potentially imply – in this case for central institutions such as the family and marriage – that is evoked. Moral indignation alongside promises of hope and a better life both play a role in the constitution of a public opinion and how these practices are legitimated. It is therefore of interest to trace the contours of the early disagreements and the sounding board against which these resonated. These factors have a bearing on understanding future controversies regarding the legislation of biotechnology.

The different positions from which the discourses emanate disclose some core values about family, kinship and filiation as well as the meanings attributed to them at the time. Central among these is the importance attributed to marriage (and its purpose) and the varying significance, for men and women, of *not* having a child. In this regard it is possible to discern a tension between biological principles of identity and belonging (especially for women) and social ones (for men, expressed through both adoption and AID). Thus this retrospective presentation of a historically specific event makes it possible to discern the shifting grounds for evaluating assisted conception and hence the gradual change in values that these processes entail (including marriage, the parent–child relation, the best interests of the child and children's rights). Retrospectively, it might even be possible to suggest that these first public debates initiated a process of naturalizing technologies of assisted conception, or as one of the committee members said: they contribute towards its demystification. This process occurs at the interface between existing and possible practices and the publicity and public interest surrounding them, whether this takes the form of sensationalist views or the form of documented medical practice, or both.

As mentioned, the 1953 parliamentary report did not result in a proposal to Parliament and no legislation was passed.[30] I have not been able to discover what the reasons for this inaction were. When I asked legislators (and bureaucrats involved in formulating later laws) to offer an opinion, they were either not aware of this earlier attempt or they had no grounded opinion.[31] However, one medical doctor (who had also been involved in the later legislative processes) suggested that it could have had to do with the fact that the government of the day (Labour) did not want to propose a law that they were not sure would pass through Parliament. This political explanation seems plausible, especially when taking into consideration that the same reason has been given for why later announced revisions of the law have not been formulated as planned. Proposing a new law (or important revisions of an existing law) requires a certain overall consensus, ensuring that the necessary majority will prevail. If not, there is no point in putting the proposal before Parliament. Thus timing is of the essence. But as we shall see, in the case of the laws pertaining to assisted conception and later biotechnology, consensus is precarious, especially if members of Parliament are freed from their party loyalty and allowed to vote according to their conscience.

The Making of a Law

The first Norwegian IVF baby was born in 1984 at the Regional Hospital in Trondheim, under the direction of Kåre Molne and Jarle Khan, two pioneering infertility medical doctors in Norway.[32] It was this event that prompted the politicians to act, lest matters get out of hand. Among members of Parliament as well as the medical profession and the public, there was a general agreement about the need to regulate. More than thirty years had passed since the failed attempt to legislate AID, and in the intervening years (especially in the 1970s) different processes supporting such a move had got underway. During this period the abortion law granting women self-determination was passed; cohabitation (rather than marriage) was becoming more and more common; single mothers were no longer stigmatized; and women were entering higher education and the labour market.

In 1987, Parliament passed the Artificial Procreation Act.[33] The Act was based on a precautionary principle that was explicitly iterated and subsumed under an overarching regard for ethics. Two overriding concerns were expressed regarding the aim of the legislation: to secure the best interests of the child, and to ensure the interests of society by hindering selective human breeding and the commercialization of reproduction.[34] Thus two significant elements are introduced: one articulates concern for the child; the other, concern for practices that violate human dignity.

As already mentioned, there was a general consensus about the need to legislate. However, it does not follow from such a consensus that the regulations necessarily be restrictive. Moreover, it is an open question whether the restrictions that were put in place can be understood in terms of the nature (sic) of the issues at hand – that is, that egg donation fundamentally breaks with Norwegian notions of motherhood – or whether the restrictive law reflects a very guarded attitude to new technologies and developments in science.[35] In fact, both (and other) concerns were articulated in the debates, and I will broach some of them. However, issues pertaining to understandings of kinship and relatedness (such as the differential treatment of sperm and egg donation) will be dealt with extensively in the next chapter. In brief, this first law stipulates the following:[36]

> Chapter 1: Artificial procreation can only be carried out by institutions authorized by the Ministry, and only these said institutions are permitted to freeze sperm. The import of sperm is only allowed by special permission from the Ministry. The freezing of unfertilized eggs is not allowed. Fertilized eggs[37] can only be used for reinsertion in the woman from whom the eggs were taken and cannot be stored for more than twelve months. Research on fertilized eggs is not permitted. Artificial procreation is only allowed for married women, and consent from the woman and her husband is required.

Chapter 2: Artificial insemination is only allowed when the husband is infertile or the carrier of a genetically inherited disease. The medical doctor chooses the sperm donor. The sperm donor's identity is to be kept secret. A sperm donor cannot be given information about the couple or the child's identity.

Chapter 3: Conception outside the body can only occur when the woman is unable to conceive. Such treatment can only occur with the couple's own egg and sperm cells. Fertilized eggs can only be returned to the woman from whom the eggs were taken.

Thus, the Act restricts assisted conception to married couples; it permits AID with anonymous donor sperm; it prohibits egg donation by insisting that only the couple's own gametes may be used; and it prohibits research on fertilized eggs. It restricts the practice of assisted conception to authorized institutions; and it gives the practising medical doctor the authority to decide whether treatment is to be initiated and to choose a suitable sperm donor.

Most of the politicians involved in the formulation of the 1987 law stated that the making of it was in many ways an exceptional situation (Melhuus 2005).[38] There are several elements that contribute towards this qualification. This was a new law, being the first time assisted conception was to be regulated in Norway. This fact made the process unique and significant in the minds of the legislators. One even characterized the event as 'making history'. The field that was to be regulated was one that on the whole people were ignorant of, including politicians. This very ignorance contributed to the process in many ways. The practices that were to be subject to law were vital issues, involving ethical questions. They were considered to be beyond the ordinary realm of politics.

The fact that IVF had become a reality in Norway and that Norwegian doctors had succeeded in producing an IVF baby made it evident that reproductive technologies were developing at a pace that demanded legislation. Most of the politicians I talked to reflected this attitude: a fear concerning the implications of the technologies; and the uncertainties involved and hence the need for control. The general feeling that I gleaned from the interviews was that the potentialities of these technologies are awesome, for some even frightening. Not only did they concern basic notions about human beings and practices that involved fundamental life processes, but they also evoked a sense of the inevitable and irreversible: that once processes (of this kind) were set in motion, they are almost impossible to reverse. As one politician stated: 'Biotechnology concerns central ethical questions … Once you start you cannot change the course. It is this quality of being irremediable … We cannot afford to make a mistake'. Another person stressed the importance of creating a public awareness about where to draw the line. As he said: 'Research moves the limits; politicians have to set them. It is the duty of Parliament to generate [public] debates'. Yet another compared his position (as a politician who has to decide on assisted conception) to being a witness to someone committing suicide: 'You have the right and the obligation to intervene'. Moreover, there was recognition that there was something special about regulating things as restrictively as was decided upon.[39] Several of those interviewed alluded to Norway being a 'different kind of country' (*et annerledes land*); some meant it ironically, others gave this 'difference' a positive value. 'We do what we think is best (for us), in spite of what others may think', is the sense conveyed. As a person in favour of the restrictions said: 'We are a bit special in Norway in the sense that it is very regulated and [we have] very strict provisions', the point being that to be 'a bit special' is all right. In fact, the attitude that Parliament and government have an obligation to act on behalf of society on these matters was a salient feature of the reflections of all the politicians I spoke to, no matter what their position on each specific issue. (So much so, that applying a precautionary principle in these matters appeared self-evident.)

A major concern of the debates, and which was reiterated in my interviews with the politicians, was the question of knowledge and competence building: What do we (really) know about these technologies and their effects, and who should be able to acquire such knowledge? Granted that there was a general feeling of ignorance about the whole field of biotechnology, the following comment by one of the politicians is intriguing: 'The law is not based on expert knowledge – that is what makes it so special'. This statement highlights a tension between expert knowledge and lay knowledge, not least the fact that this law was seen as grounded in other values than those represented by the so-called experts (that is scientists and/or medical doctors). The comment underscores the significance of lay opinions in two senses: One has to do with personal convictions – or conscience – where matters regarding essential life processes are in the last instance a question of *livssyn* (a Norwegian word which alludes to belief, but not necessarily of a confessional kind). In other words, matters of vital concern cannot be reduced to 'ordinary politics'; they rest ultimately on each person's individual values of what is right (whether these are grounded in religion or not). The other meaning of 'lay' has to do with common sense – that in these matters, a lay person is as entitled to an informed opinion as much as, if not more so, than an expert in the field. You do not need to be an expert in the field in order to know what is right.[40] This attitude was reflected in the questions related to knowledge and competence building on two counts: it challenged the authority of scientific experts; and it placed biotechnology within a moral universe, understood as being beyond the ordinary realm of politics. However, the opinions of the involuntary childless (who were of course 'heard') were not given much weight, despite the fact that their condition was recognized as both extremely difficult and painful.

Overall there is perceived to be a problem concerning rapid developments within science and technology and that it is impossible to keep pace with these new developments. Moreover, there was a prevailing attitude that it is in the nature of science (and scientists) whereby, if left to their own devices, scientists will always push to supersede existing limits. Therefore, one cannot leave it to the individual practitioner – or scientist – to decide what is right. Opinions on these matters reflect a view of the relationship between 'research' (in the field of biotechnology) and government.[41] There were those who stated that it is the job of politicians to be ahead of research – in the sense that government can and should stop questionable or unethical developments before they are a reality. As one put it: 'We must not go faster ahead than what we are able to control. The way it is now, legislation is adapting to research'. Or, as another person said: 'Things happen that you never have thought about until Dolly is born, so you have to try to be a horse's length ahead and it may seem then that we are restrictive'. Others scorned this view, saying that you cannot hinder – or regulate for – something if you do not know what it is. For those who were positive about the uses of biotechnology, the attitude was: 'it is not desirable nor possible that law makers can be ahead of technological developments; one has to regulate in their wake'. Of the two camps – those who are pro-technology and pro-research and those who are far more sceptical – the sceptics won. The restrictive laws were enforced – and reinforced in later revisions.[42]

The question of competence building has another side, but reflects some of the same concerns. At issue was (and still is) whether Norway should be part of an international community of researchers and partake in the international development of science and technology within the field of biotechnology, including assisted conception. By forbidding research on embryos, this possibility is in part precluded in that Norwegian researchers in this field will have to travel abroad in order to gain knowledge of and practice in new methods and developments. Thus, medical doctors have argued that it is hypocritical and parasitical for Norway to uphold a restrictive law while at the same time making use of knowledge and technologies developed elsewhere (e.g., Hazekamp and Hamberger 1999). This, they said, involves 'exporting our ethical dilemmas'. This view was also reiterated by the government in 1993, in conjunction with the revision of the law.[43] The counter-argument is (and was) that Norway has the right and the obligation to practice what Norway – or Norwegians – deem ethically correct. The problem of course is that there is no agreement on what the ethically correct position is. Hence also the continuous revisions of the law. Nevertheless, when it comes to the question of ethics it is as if those who are more restrictive perceive themselves as 'more ethical' than those who are not. So much so that in the parliamentary debate those supporting the more liberal aspects of the law found it necessary to state explicitly that their positions are also ethically grounded.[44]

The government responsible for putting the proposed 1987 law forward was run by the Labour Party, which has on the whole been positive towards the introduction of biotechnologies and their application. There was intense contact between the Minister of Social Affairs, the members of the Committee of Social Affairs and even the prime minister (who at the time was Gro Harlem Brundtland) in formulating and finding support for the law. Members of the committee that I talked to not only felt that they had a special responsibility but also that there was a real possibility of influencing the formulation of the law. In fact, when I questioned them about their impact on the law, many explicitly said that they had had a direct influence. This influence had different channels: through discussions in the committee itself and through discussions with their respective party fractions, winning co-members over to their point of view. The degree of influence also had to do with the amount of knowledge each representative felt they had. Several of the representatives expressed frustration at not knowing enough – whilst also realizing that the little knowledge they had was more than co-members of their respective parties had. Some members established permanent reference groups with necessary expertise; others called upon experts when necessary. Moreover, and as mentioned, there was also an explicit recognition of the ethical nature of the issues to be treated. This was reflected in the final vote. Members were allowed to vote on the basis of their personal conscience, their *livssyn*, and were not obliged to vote along party lines. This is unusual, as is the fact that two parties voted against the law *in toto* (the Christian Democrats and the Centre Party). Moreover, the voting record indicates that the margins were in some cases small.[45] All those I talked to were unanimous in proclaiming that this was an exceptional situation, a fact also confirmed by the official records.

There is another point to be made. Although I have stressed the fact that the issues treated in this law were considered to be out of the ordinary, there was nevertheless some pragmatic politicking going on too. This had to do with two controversial issues: the rights of private clinics to provide IVF treatment and the legalization of research on embryos. The Labour government had originally proposed legalizing limited research on embryos (in order to improve IVF methods). It had also proposed limiting IVF treatment to public clinics only. For the Labour Party it was important to curtail the expansion of private medical care, in contrast to the aims of the opposition. At the time, there was one private clinic already offering IVF treatment, thereby openly challenging the public health-care system and its potential monopoly on practices of assisted conception. The question of private clinics, however, was not just debated in terms of who was to be allowed to practise; it was also a matter of whether the competence and expertise involved in developing these techniques should be primarily based within public hospitals and hence part of the public health-care system. These then are questions that first and foremost have to do with the public health-care system and its role.[46] However, research on fertilized eggs is a different matter. This has to do with the moral status of the embryo.

Initially, the Conservative Party had supported the government's move to allow research on fertilized eggs. However, the Conservatives shifted their position in a compromise with the Christian Democrats: the Conservative Party agreed to vote against research on embryos and the Christian Democrats agreed to vote for authorizing private clinics to offer IVF treatments.[47] This compromise gave a majority vote on both counts. Regarding research on embryos, the vote was close: 48 voted against permitting such research, whereas 46 voted for (the Labour Party and the Socialist Left Party voted for the proposal to permit embryo research, with the latter party changing its position with the later revisions). So, although the issues were framed as ethical and a matter of conscience, there was nevertheless room for horse trading. Moreover, as the voting record indicates, an issue deemed 'most ethical' (by some) is also one that is most contested. It brings home a point previously made: there is no consensus on these matters in Norway.

This was evidenced in three other issues: the use of anonymous sperm, the freezing of embryos and the limiting of treatment to married couples. Although there was a majority for upholding the anonymity clause (53 to 42), there were voices in the Centre Party and Socialist Left Party insisting that it is the right of a child to know its biological origins. There were also Christian Democrats who were against the use of donor sperm entirely, arguing that this practice would split the biological unity between mother, father and child by introducing a third party. Freezing embryos (for 12 months) was also a matter of concern (although the majority for the measure held a clear margin: 62 to 31). For example, the Christian Democrats state: 'The freezing ... of embryos ... is artificial, unnatural and an undignified intervention in the development of a human life'.[48] Regarding who is to have the right to treatment, the government lost its move to extend treatment to cohabiting couples (of opposite sexes) by four votes. Thus, marriage was still upheld as the central procreative institution. However, there was general agreement that to have children is not a right per se, though there have been disagreements about what priority to give infertility treatment within the public health-care system.[49] At issue here is the classification of infertility (as an illness or not) and the principle of equal access to health care services. In other words, the question of permitting private clinics to offer IVF treatment (and to give such treatment low priority in the public health-care system) is also a question of who will then be able to benefit from these services – that is, those with the means to pay for it. Such a position goes against an ethos of equal access to health care services irrespective of economic means. This issue is reiterated with regard to those who can afford to travel abroad for treatments that are not permitted in Norway.

In the wake of this debate and subsequent vote, Parliament requested that the government submit a White Paper (*stortingsmelding*) on biotechnology that could form the basis of an informed ethical debate on further guidelines for the use and application of biotechnology in relation to humans. The Ethics Committee was appointed in 1988 and submitted its report in 1990.[50] The report from the Ethics Committee subsequently laid the ground for the White Paper presented to Parliament in 1993 on biotechnology relating to human beings,[51] which formed the basis of the 1994 law. In the meantime, the Norwegian Biotechnology Advisory Board was appointed in 1991, under the Ministry of Health and Social Affairs.

The First Revision, 1994

In March 1994, the Ministry of Health and Social Affairs presented to Parliament a bill relating to the application of biotechnology in medicine. This bill was based on a substantial report entitled 'Biotechnology Related to Human Beings'.[52] The proposal was debated in Parliament and passed in June 1994. With the bringing of the Act Relating to the Application of Biotechnology in Medicine into law the following August,[53] the Artificial Procreation Act of 1987 was repealed. Assisted conception was now placed within a broader category and regulated under what was termed biotechnology and medicine. This allowed for the regulation, within one and the same Act, of such practices as research on embryos, preimplantation diagnosis, and later cloning, prenatal diagnosis, genetic testing after birth, and gene therapy. Hence, this act is significantly broader in scope than that of 1987. Moreover, and in contrast to the Act of 1987, the later Act includes a preamble which clearly states its aims:

> The purpose of this Act is to ensure that the application of biotechnology in medicine is utilized in the best interests of human beings in a society where everyone plays a role and is fully valued.[54] This shall take place in accordance with the principles of respect for human dignity, human rights and personal integrity and without discrimination on the basis of genetic background, on the basis of ethical norms relating to our Western cultural heritage.

This preamble is significant as it is indicative of some of the ethical issues that underpinned the debates and issues that were raised in conjunction with the Act. In fact, even the preamble itself was subject to discussion. Although there was a consensus that the law needed to be set within an ethical framework, there was disagreement on the actual framing. All were agreed that the law must be based on the fundamental idea of human dignity – that all humans have the same value (thus stressing the inclusion of the disabled) – and on an idea of society based on the value of solidarity ('where there is room for all', a phrase that was added as a consequence of the debates in Parliament). An important issue of contention was whether the central value reference should be 'our Christian heritage' rather than 'Western'. The main objection here was that there should not be a Christian object clause with regards to the application of biotechnology.

The regulations proposed by the Labour government were on the whole more liberal than the 1987 Act, although with regard to assisted conception some proposals were adjustments rather than radical changes. For example, there was a proposal to extend the indications for allowing IVF treatment to include male and inexplicable infertility. There were also proposals to allow IVF with donor sperm, which was not permitted earlier, as the combined technologies made for 'more manipulation',[55] and to extend the period of storage for embryos from one to three years. The government proposed that assisted conception be extended to women living in stable partnerships with men (that is, not restrict treatment to married couples), and gained a majority for that move. The government also put forward a provision permitting research on embryos under specific conditions, which was rejected. Significantly, the government did not even propose permitting egg donation.[56]

The question of anonymity was debated in Parliament and several members argued for rescinding the clause with reference to a person's natural desire to know its biological origin and a child's right to know where it comes from. Those arguing against stated that rescinding the anonymity clause would be tantamount to a 'social experiment', whose consequences are difficult to foresee. To give a child the right to know a father who never wished to be that child's father was just not tenable. An additional argument for upholding the anonymity clause had to do with the availability of sperm. In 1994, Norway imported most of the sperm used for donation from a sperm bank in Denmark.[57] By rescinding the anonymity clause, this would no longer be possible (as sperm donation in Denmark is anonymous) and, hence, the very offer of AID would not be available in Norway. Those in want of anonymous sperm would then have to travel abroad, and perhaps obtain treatment not regulated by any authority. Moreover, they would have to pay for it themselves.

The government gained a majority for most of its proposals, with shifting constellations of individual members' votes. However, it did not gain support for permitting IVF with the use of donor sperm. Although there were strong opinions against upholding the anonymity of sperm donation, this provision was upheld, as was the prohibition on egg donation and the ban on research on embryos. In addition to the provisions regarding assisted conception (covered in § 2) and research on embryos (§ 3), regulations concerning the following were introduced: preimplantation diagnosis (§ 4), prenatal diagnosis (§ 5), genetic testing after birth (§ 6), gene therapy (§ 7) and finally general provisions (§ 8). With the consistent support of the Conservative Party and the Progress Party, the government gained support for the rest of its provisions, though there was ample resistance. In the case of preimplantation diagnosis (PGD) – that is, genetic examination of the fertilized egg – the minority put forth an alternative proposal, forbidding PGD. And in the case of prenatal diagnosis (PND), the discussion turned on the issue of disabled people, society's attitudes to disability and the question of abortion (and the need to reduce the number of abortions). In both cases the crux of the matter was the possible elimination of an undesirable or impaired fetus (a term deliberately used by the Christian Democrats, to cover both embryo and fetus). Nevertheless, these provisions in the Act are on the whole fairly restrictive (to be applied in special cases only) and practical implementation requires authorization from the Ministry of Health and Social Affairs.

According to Sirnes (1997: 223), the arguments and perspectives put forth in 1987 and those leading up to and culminating in the 1994 legislation were more or less the same. Significant new elements are not introduced, despite the interval of several years, the number of documents produced and developments within the biotechnology field. Adding to this the fact that the political constellation had not changed substantially (Labour was still in power), it is perhaps not so surprising that the 1994 legislation did not change radically from that of 1987. The law was still restrictive on the most controversial issues, but had liberalized some practices (for example, extending the period for freezing and storing embryos) which, more than anything, were adjustments to an existing practice and more pragmatic than matters of principle. This was to change, however, with the revisions of 2003. There is a change of winds, favouring those whose attitudes to biotechnologies are more restrictive.

The Second and Third Revisions, 2003 and 2007

In 2001 a coalition government involving the Christian Democratic, Conservative and Liberal parties came to power. The prime minister – Kjell Magne Bondevik – was from the Christian Democratic Party, as was the Minister of Health, Dagfinn Høybråten.[58] There is no doubt that it was important for the Christian Democrats to influence fundamental health policy in Norway. Thus, in November 2003 the government introduced a revised proposal of the Biotechnology Act of 1994. In fact, Parliament had already decided in 1994, when the Act was passed, that it was to be evaluated after five years.[59] However, it took a change in government to fulfil this obligation.

The proposals put forth by the new government were based on a White Paper submitted to Parliament in 2002.[60] This was debated in Parliament in June of the same year,[61] and the revised Act was passed in Parliament in November 2003 by 67 votes to 2.[62] However, although at the final stage there was an almost unanimous vote, this consensus conceals the fact there were some highly contested issues. In general the majority consisted of the coalition parties – the Conservatives, Christian Democrats and Liberals – with support from the Socialist Left Party and the Centre Party. Together they upheld the more restrictive provisions. The Labour Party and the Progress Party put forward more liberal proposals but were defeated.[63]

The most radical change that the majority proposed was the repealing of the anonymity clause with regard to sperm donation. The Labour Party, wanting to uphold the anonymity clause, presented an alternative proposal, but this was defeated. Thus, a practice, contested since 1953, was overturned. The main argument for revoking the anonymity of sperm donors was a child's right to know its biological origin (*opphav* is the Norwegian term often used, which connotes descent). In fact, the majority argued that 'a legal precondition (for assisted conception) is that biological origin can be clearly defined as one biological mother and one biological father'.[64] This argument also supports the continued prohibition on egg donation. Thus a fundamental link between identity, kinship and biogenetic origin is established. The concerns about origin and truth iterated in the 1950s are finally put to rest. Moreover, framing the question of anonymous donor sperm in terms of children's rights – and the best interest of the child – has surely been constitutive of the more general change in opinion on this matter.[65]

When I asked politicians whether they had, in the course of time, changed their opinion on any of the issues involved in biotechnology law, anonymous sperm donation was invariably mentioned. If they were to vote today, they said – and keep in mind that my conversations with politicians took place in 1999 and 2000 – they would vote for the rescinding of donor anonymity. References were made to the adoption law and to international conventions, stating that a child has the right to know its biological parents.[66] In their view, donor children and adopted children are 'the same', and hence have the same rights. There should be no differential treatment of children – all children have the right to know their biological origin, and reference was made to Article 7 of the UN Convention on the Rights of the Child (ratified by Norway in 1991). These are exactly the same arguments the majority put forth in their support of known donors.[67] The fact that adoption and assisted conception are two very different ways of creating parenthood is irrelevant.[68] Moreover, arguments put forth by medical experts about the problem of recruiting donors and the suggestion that AID would likely be suspended in Norway carried little or no weight. On the contrary, the majority argued that the use of known donors might ensure more responsible donors, by the mere fact that they know they might at some later date be confronted with a 'donor child'.[69] The new law specifies that a child born through assisted conception with the use of donor sperm is, at the age of 18, entitled to information about the donor's identity. (However, the law does not oblige parents to inform the child of the method of conception.) A donor register is to be established; specification about the donor and his rights are given; and the Ministry of Health is granted the right to regulate the establishment of sperm banks.[70]

On the whole, the law upholds the earlier restrictions – restricting assisted conception to heterosexual couples, prohibiting research on embryos, prohibiting egg donation, and restricting the application of PND and PGD. Some adjustments were also made. The storing of fertilized eggs is extended to five years and egg cells may under certain circumstances also be stored.[71] The law also states that the couple should be informed about the medical and legal implications of treatment, and they must also be informed about adoption.[72] In other words, medical doctors are now required to inform patients about adoption as a possible alternative when informing them about infertility treatment. The law now permits the use of donor sperm in combination with IVF. One small but nevertheless significant point is that, with this Act, it is suggested that the term 'artificial conception' (used systematically in the White Paper) be dropped in favour of 'assisted conception'. This represents one small step to 'naturalize' these procreative acts.[73]

The debates in Parliament echo some of the tensions and tendencies mentioned earlier. These have to do with ethics and ethical dilemmas, attitudes to knowledge and science, and the role of politics and legislature in intervening in the form of regulations. Two issues in particular highlighted these tensions. One was in conjunction with the continued prohibition on research on fertilized eggs. The other had to do with the restrictions applied to prenatal diagnosis (PND) and preimplantation diagnosis (PGD). Both these technologies allow for the sorting away of fetuses or fertilized eggs on the basis of certain characteristics. In both cases, the debate was heated, in an effort to appear 'most ethical'. The opposition did not like insinuations that those with more liberal attitudes to technology and science had the less ethical position – and hence by implication the mere fact of being more restrictive implied by definition a more ethical position. In the discussions, the precautionary principle is held up against the fear of new knowledge.[74] Protection of the most vulnerable point of life – life at its inception – and respect for human dignity were held up as arguments against research on embryos; the possible sorting of life on the basis of unwanted characteristics was held up against both PND and PGD. Born life was held up against unborn life; selective abortion against abortion on demand; and a woman's right to decide when and if to make use of ultrasound was set against the fear of potentially laying the grounds for a sorting society. (These issues are elaborated and discussed in Chapter 5.) In other words, at issue were fundamental biopolitical questions concerning the obligation and responsibility of the state (through its legislative practices) vis-à-vis society in vital matters, such as definitions and meanings of life, procreation and choice. In which direction should Norwegian society go? Where should the line be drawn?

In 2005, a new coalition government composed of the Labour, Socialist Left and Centre parties came to power. In 2007, this majority government presented proposed revisions to the Biotechnology Act before Parliament.[75] This revision contains two new proposals: to permit research on supernumerary embryos (under certain specifications) and to permit preimplantation diagnosis under certain circumstances. Both proposals were passed with a clear majority. The coalition government had support from the Progress Party, while the Conservatives and Christian Democrats voted against the law *in toto*. Interestingly, the government did not propose permitting egg donation, despite the fact that the Labour Party had in earlier debates signalled its positive attitude towards it. This indicates not only that there was no majority to be had on this question, but also that the Labour Party itself might be split on the issue. Thus, while other issues were shifting ground – for example, the Socialist Left Party changed their view on embryo research – the prohibition on egg donation stood. With this revision the Labour Party finally won a majority for embryo research, an issue that had been on their agenda since 1994.

Research on embryos was, on the one hand, debated in relation to stem cell research. The former coalition government (being against research on embryos), had given priority to research on adult stem cells. The present government, however, stressed the future potential of embryonic stem cells in the cure of serious diseases, though they were accused of 'overselling' their bid. On the other hand, the debate turned on the so-called 'supernumerary' embryos – that is, those that would be destroyed due to being no longer needed in cases of IVF treatment. With regards to the embryo, the basic divide was between those who saw the embryo as insipient life and in need of protection and those who argued (albeit uneasily) that it is better to permit research on embryos than let them go to waste. Thus the issue not only turned on the moral status of the embryo but also on whether destruction of an embryo is ethically more acceptable than research that might eventually be life saving. It is therefore interesting to note that the question of embryo destruction as such was not debated. With regard to PGD, the debate turned on selection and the possibility of consciously sorting away those that are different. In this connection, the notion of *likeverd* (equality as equal worth) was stressed.

The following quotes capture the mood of the opposition. According to a representative of the Conservative Party, the then proposed revision 'shifts important ethical limits … as many of the issues to be debated are about one fundamental ethical question: who has the right to life?'[76] And she continues: 'The Government has proposed – and gained a majority – to permit research on supernumerary embryos. For the Conservative Party this is ethically unacceptable as it implies reducing incipient life (*spirende liv*) to a tool for research … [T]his is an instrumentalization of human life, violating human dignity in the name of technological progress'. The articulated fear is that incipient life will become a commodity. Or, in the words of a Christian Democrat:

> The Christian Democrats support biotechnological research. It gives us important knowledge, but it has to be anchored ethically. Through research on fertilized eggs we can gain knowledge about serious diseases and better methods for genetic examinations. However, this may in turn lead to increased selection. Through the new knowledge that we obtain as a consequence of research on fertilized eggs, more fertilized eggs with an illness can be sorted away.

With regards to PGD, the following statement from another member of the Christian Democrats is symptomatic:

> At the same time, I feel that words like 'sorting society' are correct in this connection, because the proposed legislation makes it possible for us to make conscious choices and choose to eliminate that which is different, that which we feel is not healthy enough … [T]he organizations of the disabled have said that we must be cautious and not create a development where we actually say that some people are different, and who will feel that they are less welcome in society. That is a perspective I find very important: that we create a society where all are welcome, where all have the same worth.

Again the question of selection was raised and the problem of choosing to 'sort away' that which is – or those who are – different. The unease here is that the law potentially opens up the possibility of a 'sorting society', where persons (read fertilized eggs) with particular characteristics will be systematically eliminated. This was voiced both in connection with embryo research and in connection with the revisions pertaining to PGD. Both sides in the debate claimed that their position was in accordance with the preamble: respecting human dignity and recognizing the utility of biotechnology in a society where everyone plays a role and is fully valued. Yet, the majority is accused of furthering perfection – the flawless (*lytefri*) person – and hence a maximization of utility, against which stands a seriously ill person, who is neither perfect nor useful. At issue is whether embryo research and PGD reduces human beings to means to an end rather than an end in themselves, and the fear and apprehension that this area of biotechnology marks the first steps on a slippery slope.

In Sum

This chapter has covered a legislative process that commenced in the 1950s with the first attempts to legislate AID and ends with the revision of the Biotechnology Act in 2007. During this time span of over fifty years, developments in reproductive technologies have been dramatic, radically changing practices and perceptions of procreation in Norway as elsewhere in the world. The Norwegian legislators were not regulating in a vacuum. They were influenced by what was going on in their own society as well as what was happening elsewhere. My intention has been to document the Norwegian legislative responses to developments within biotechnologies and assisted conception, and demonstrate the working relations between biotechnology and law. As we have seen, there has been an overall consensus regarding the need to regulate the application of these technologies, but there has also been considerable disagreement about the extent of the regulations. Thus, shifting political constellations have determined the outcome of the drawing up of legal regulations. Despite the fact that research on supernumerary embryos (one important point of contention) is now permitted, the Norwegian legislation remains restrictive. The consistent prohibition on egg donation is but one example. The restrictions regarding the application of PGD and PND are others. The next two chapters focus on these issues. Chapter 4 is an attempt to come to grips with the continued resistance to egg donation. Reading the law as a negotiation of kinship relations, I explore meanings attached to sperm and eggs and what they reveal about notions of maternity, paternity and filiation. Chapter 5 takes a broader view, engaging in a discussion of biopolitics through the notion of the sorting society. Here I present yet another problem of conception that revolves around the moral status of the embryo and the question of selective breeding.

Chapter 4

THE INVIOLABILITY OF MOTHERHOOD

> Until recently, difference between male and female parent rested in the two ways in which procreative parenthood was traditionally established, mother through birth and body issue; father through presumed coitus with mother. These two routes to knowledge supposed two kinds of linkages … The mother's biological tie to the child is at the root of her social obligation to it; the father's social tie to the mother is at the root of his knowledge of biological connection with the child. In other words, in pre-technology Euro-American systems, motherhood and fatherhood required different kinds of proof to establish the facts of procreation. And fatherhood appeared to require more proof than motherhood.
> —Marilyn Strathern, *Property, Substance and Effect*

> Norway the Best Mother-country in the World.
> Norway is the world's best country for mothers, whereas the USA is far down the list, shows a ranking made by Save the Children.
> —*Verdens Gang*, 4 May 2010[1]

Establishing Parenthood

At the core of Norwegian understandings of kinship, as in Europe more generally, is the idea that 'blood is thicker than water'. Blood has been – and to some degree still is – the metaphor used to designate the biological foundation of kinship relations and to indicate that it is the sharing of substance that makes people kin. To some extent in today's lay understandings, notions of blood, biology and genes are interchangeable: they all point to some natural, substantial aspect of kinship which is assumed to be the foundation of kin relationships. However, it is an empirical question to what extent blood and genes are conflated. As Edwards states, 'genetic connection is not [necessarily] coterminous with genealogical connection, and "blood" does not fully map onto genetics, just as it does not fully map onto biology' (Edwards 2009: 10). Whatever the case, it seems that European kinship ideologies are still grounded in what is perceived as some kind of 'natural' foundation. Such particular perceptions are the subject of this chapter. More specifically I explore the workings of a link between biology – or biogenetics – and identity to establish certainty in kinship relationships tied to filiation.

Though blood relations have been a precondition for establishing kin relations, they are, as we know, not sufficient for establishing legitimate kin relations. Marriage has been – and to a certain degree still is – the institution which has endorsed legitimate kinship relations. Nevertheless, in Norway as elsewhere, significant changes have taken place. Whereas in the past (indeed in the 1950s) children born out of wedlock were considered illegitimate (the term for an illegitimate child – *uekte barn* – connoted that it was not 'authentic'), and unwed mothers were stigmatized, this is no longer the case. Legitimate procreation is no longer regulated through such moral categories. In fact, in Norway the distinction between legitimate and illegitimate children no longer carries any meaning; the categories as such are obsolete. The fact that this distinction is no longer culturally relevant can be seen as an expression of the shifts that have occurred in both the perception and organization of relations between men and women. In many ways this loss of meaning sums up a series of other socio-cultural changes that have occurred in Norwegian society. It is a nexus around which the relations between sexuality, procreation and morality turn. It is also a precondition for the socio-cultural acceptance of assisted conception, capturing a shift in concern from sex without procreation to procreation without sex.

Today, both cohabitation and divorce are common, so much so that they form part of the expected life cycle (Lappegård 2007). Moreover, reconstituted families are an accepted, if not unproblematic, feature of everyday lives – for parents and children, not to mention grandparents (Jørgensen 2001). Until very recently, marriage has been by definition heterosexual. In 2008 (with effect from 2009) Norway amended the Marriage Act of 1991, granting same-sex couples the right to marry, the right to adopt, and – for lesbian couples – the right to assisted conception.[2] In different ways, these new family forms challenge traditional understandings of what a family is and should be, changing the meaning of marriage, the relation between spouses and the relations between parents and children. These are in turn exacerbated by the possibilities that reproductive technologies entail. These combined processes converge, highlighting a shift in the meanings of procreation and filiation.

There are two principles that underpin European notions of kinship which, in my opinion, still have significant ramifications. These principles serve me well in elucidating the particularities of the Norwegian situation as they capture some underlying attitudes to motherhood and fatherhood. The two principles are, first: *pater est quem nuptiae demonstrant* ('the father is established through marriage'); and second: *mater semper certa est* ('the mother is always certain'). The first principle is dependent on marriage. Recognizing the uncertainty of paternity, it stresses the significance of the conjugal tie. The so-called *pater est* rule has been the basis for establishing legal paternity. The second principle, however, is perceived as a biological fact. No matter what the birth conditions of a child are (whether it is legitimate or not), the mother is assumed to be known. Thus, until 1997 (with effect from 1998) Norway has not had a specific clause defining maternity. At that time, however, and as a result of the debates about reproductive technologies, and the potential fragmentation of motherhood, the following provision was added to the Children's Act: 'The woman who gives birth to the child is considered its mother'.[3]

These two principles, although not always or even necessarily articulated, have guided practices that have to do with filiation. And as we shall see, they are also relevant for my further discussion on the prohibition of egg donation in Norway. These principles rest on what have been seen as undisputable biological facts: that paternity is always uncertain and that there is no doubt about who the mother is. Although the first principle insists on marriage as the institution that establishes legal paternity, the second insists on the certainty of the mother. Egg donation, however, challenges the latter, and the practice of anonymous sperm donation is in a sense inscribed in the former. For all intents and purposes, anonymous sperm donation collapses social and biological paternity into one father, the assumption being that the knowledge of donation is considered a private matter, kept secret and not revealed. Sperm donation with a known donor makes the distinction between genitor and pater public. This distinction is embodied in the very conception of the child.

In Chapter 2, I introduced the involuntary childless and their ideas about what it is that makes a child one's own – that is, their notion of an 'own child'. At issue is the significance of shared substance in the making of kin, and this in turn has to do with the kinds of relations they envision as important. Two points seemed to me salient: on the one hand, the significance of creating a child that is in some ways understood as equally related to both parents – and it is not necessarily shared substance that makes for this equality. On the other hand, there is a biological model, based on an understanding of equal contributions of genetic material from both parents, which nevertheless represents their main frame of reference. So much so, that the absence of shared substance – as in the case of adoption – may be for some as forceful an argument as the presence of own egg or sperm. In such instances, the child works to ensure a balance between the parents in terms of their relation to the child, thereby strengthening the conjugal relation and the unity of the family. It is also important to recall one more point: that for those of the involuntary childless to whom I talked, in the process of coming to terms with their condition and the options available, the overriding concern is the desire to have a child. Thus, many are willing to circumvent the law in order to realize this dream.

The previous chapter examined in some detail the legislative processes that have addressed assisted conception and reproductive technologies, indicating both attitudes to these technologies and some of the shifts that have occurred in attitudes over time. In this chapter I remain in the public domain but with a deliberately narrow focus. My attention is primarily limited to the differential treatment of sperm and egg donation as these are articulated in public discourse and the ideas and values about relatedness and belonging that are transmitted through these. If you recall: anonymous sperm donation has been tacitly permitted since the 1930s, explicitly since 1987, but was prohibited in 2003; meanwhile, egg donation has been consistently prohibited. This is in marked contrast to Spain, Denmark, France, Belgium and Great Britain – just to mention some of the countries in Europe that permit egg donation. Thus, in comparative terms, Norway represents an interesting case towards understanding resistance to this practice.

In what follows I explore some of the meanings attached to sperm and eggs, and what they reveal with regard to underlying understandings of paternity and maternity. Underpinning these understandings is an idea of nature – or respect for nature – and notions of what is natural. I also consider an exceptional case of surrogacy, as this gives an added twist to ongoing socio-cultural and legal processes. Thus I read the legislation and the concomitant debates as a public negotiation not only of kinship (Franklin 1999: 127) but also of the categories that constitute kinship relations, with an emphasis on filiation. In that light, both the uncertainty of paternity and the certainty of maternity are important elements.

The prohibition on egg donation and the rescinding of donor anonymity are not grounded in the same reasoning. However, they nevertheless reflect aspects of the same tendencies. They cohere but perhaps in a way that is not generally recognized. This coherence, I suggest, has to do with two things: on the one hand an attitude to nature – or the natural – and motherhood, which makes egg donation difficult to accept, whereas sperm donation is acceptable; and on the other hand, a public discourse which increasingly tends to biologize identities and kin relatedness in terms of rights and the best interests of the child. Underpinning these discourses, I argue, is a drive for certainty: there should be no doubt about who the 'true' mother or father is. Moreover, there is an effect: in some ways, paternity is becoming more like maternity, rather than the other way round. This is the thrust of my argument.

Of Eggs and Sperm

In a country that prides itself for its gender equality policies it is noteworthy that egg and sperm donation have not, until very recently, been framed in these terms (see Spilker and Lie 2007). Although voices have been raised pointing to the fact that the biotechnology law discriminates between infertile men and women by permitting sperm donation and not egg donation, this argument has had little purchase. This can partly be explained by the fact that large parts of the women's movement saw new reproductive technologies as a new form of patriarchal control over women's bodies (Hellum, Syse and Aasen 1990; Hellum 1993). However, this argument, important as it was at the time, is compounded by another: an idea of a 'unitary motherhood' and a notion of 'mother belonging'. Whether these notions will in turn yield to an equality discourse, only time will tell. But, there are indications that women are beginning to change their minds, and that the prohibition on egg donation will be framed in terms of gender discrimination.

It is therefore relevant to ask how withholding eggs from circulation has been made possible whereas sperm can be made to move. As biogenetic substances, both sperm and egg are equally vital for procreation to occur.[4] It would not be unreasonable to argue that it is this vital quality – creating life – that is the most essential one, and therefore it represents an adequate basis on which to argue for an equivalence between egg and sperm donation. However, this has not been the case. Being of the same essence is clearly not enough – or at least not good enough. Rather, it is what distinguishes these two substances that is emphasized. As one politician exclaimed, when confronted with the question why permit sperm donation and not egg donation: 'Egg and sperm is not a question of gender equality. It has to do with biology!'

Egg and sperm have some properties that make them qualitatively different. Sperm, in contrast to eggs, is a renewable substance. It is also (and again in contrast to eggs) easily available, visible and abundant. In some respects, sperm is more like blood, which is also considered a vital substance.[5] Eggs on the other hand are limited; they are not renewable and their quality diminishes with a woman's age. They are (for most of us) invisible, and they are (at least within Norway) non-exchangeable. Yet it is not these physical properties that determine their availability (or not) for donation; they are rarely if ever mentioned, even though the more arduous and costly process involved in harvesting eggs is recognized. Rather, it appears that eggs and sperm are embedded in different chains of meaning, and it is these chains of meaning that determine their varying significance and hence value. Eggs are, in a way sperm is not, inalienable – classified as unambiguously singular objects. They 'belong' where they come from. At least this has been a dominant understanding in Norway.

I have elsewhere discussed the differential treatment of sperm and egg donation in terms of exchange (Melhuus 2003). There, I examined the nature of these substances as objects of exchange with a view to the relations involved. These relations are circumscribed by differences in availability: whereas sperm may circulate, eggs may not. At the time the article was written, only anonymous sperm was permitted in Norway and the sperm used was imported from a sperm bank in Denmark (see Chapter 3). Anonymity works so as to create a maximum degree of distance between donor and recipient. It is in fact the sperm (and not the donor) that is the subject of this exchange.[6] Thus, anonymity is a way of precluding excessive meaning, by withholding information. The secrecy implied not only hides meanings, it is also a way of getting rid of – even anticipating – unwanted images (of, for example, 'two fathers'). In other words, ignorance is bliss. The silence surrounding the phenomenon of anonymity is in itself meaningful. It is – or at least has been –

the whole point. Not only is the fact of a donor undisclosed (and hence relations of exchange go unacknowledged) but the relation between the child and its social father appears as if 'natural' (based on biogenetic relatedness).

Rescinding donor anonymity radically shifts the terms of the exchange.[7] The introduction of a named donor changes the nature of the relationship between the involved parties. It creates relations hitherto unacknowledged, as the tacit understanding of ignorance contained in anonymity is transformed: it is drawn out of the private realm, becomes public, explicit (in that it is recorded) and hence available information. The people involved know of each other, and this creates a connection, a form of relatedness. Whether this knowledge necessarily implies getting to know each other is another question. (As mentioned, the law gives the child the right to information about the donor; it does not oblige the parents to inform the child of having been conceived by donor sperm.) Nevertheless, there is the potential for the relationships to be activated. However, it is no longer the relation between donor and recipient (be it to the woman being inseminated or the infertile man) which is significant but rather the relation between the donor and the potential child. Thus, repealing the anonymity clause shifts some meanings attached to sperm, highlighting one important aspect of paternity and filiation – that is, the biogenetic relationship.

We have seen that as early as 1953 there were voices arguing against the practice of AID (see Chapter 3). Some of these arguments were reiterated in 1987 during discussions of the proposed Artificial Procreation Act. Church representatives and Christian organizations in particular have been against this practice as, they say, it undermines the institution of marriage. One typical comment is that of the Bishop of South Hålogaland. He stated: 'This bishop must as a matter of principle reject donor insemination because it breaks fundamentally with the aspect of "one body" which is fundamental for the biblical view on marriage ... A literally "foreign body" will "break into" that which should be whole and inseparable'.[8] In the same vein, the Norwegian School of Theology (Menighetsfakultetet) stresses that AID brings a third party into the marriage as a participant in the creation of the child: 'The child will thus have two fathers, one social and legal father and one biological'. According to them, this problem has not been given due weight. They also insist that, 'based on Christian ethics it would be natural to stress the marriage relation between man and woman as both the social and biological framework for reproduction and upbringing'.[9]

However, there were also voices within the Church that based their arguments on human rights. As the Bishop of Møre stated: 'I agree with the ... committee that proposes repealing the principle of anonymity on the basis that it is a human right to know how one has been conceived, and it is important for people to know their roots. Here the parallels to adoption are clear'.[10] Likewise, the Norwegian Christian Doctors Association stated: 'In our opinion ... it should be considered a human right to be able to obtain knowledge of one's own biological origin'.[11] Such human-rights-based arguments (along with arguments referring to the best interest of the child) are voiced by many others and, as we have seen, represent the major reason for repealing the principle of anonymity.[12] It is these arguments that finally won ground in 2003.

Those who still defended anonymity stressed either the significance of social paternity or the fact that known donors would be hard to recruit. Thus the Chief Administrative Officer (*Fylkesmannen*) of Troms said: 'It cannot be said that our legislation has a major goal to assess biological paternity. This is reflected in the rule of *pater est* ... It would be more correct to say that our regulations seek to ensure that the child has a legal father, and that in that light an insemination child would not be any worse off'.[13] Or, as in the words of Kjell Bohlin, arguing the following in Parliament:

> It is strange that so many seem to place more ... weight on the biological relationship to a person who for the child is a stranger than the close relationship of a secure social father. Feelings of paternity do not come with an ejaculation. It is the warm, close relation between father and child living together which creates the bond that binds.[14]

As in the earlier debates, the arguments shift between nature and nurture. Yet, the debates from 1987 onwards reveal a marked change in the grounding of positions. Emphasis is increasingly placed on biology, but is done so by linking biogenetics to rights, identity and the best interests of the child. This tendency is reflected more generally, and gains recognition in the suggested amendments to the Children's Act regarding the ascertaining of paternity.[15] Granting men the right to have paternity ascertained unilaterally (but within certain limits) by means of a DNA test is viewed by some as a move to recognize true paternity and to rectify situations where a man has been paying child support for a child which is not 'his' (the underlying assumption being that women do not necessarily acknowledge who the 'real' father is). It is also seen as a way of getting the facts right.[16] However, others frame their reactions in terms of what is in the best interest of the child.

The Ombudsperson (*Likestillingsombudet*) for issues of discrimination expressed the following opinion:

> The Ministry's suggestion implies that in cases where there is doubt about paternity men are given the unconditional right to have their case tried. This means that men and women are given equal access to try paternity. The proposal implies that the consideration concerning the stable situation of the child (*ro rundt barnet*) is not given the same weight as in the present regulation; rather it is the significance of securing the child's right to knowledge of and possible contact with its biological father as early as possible which is stressed. *Ombudet* finds that this is a correct weighting of the various considerations to be taken in cases such as these.

However, the Centre for Equality (*Likestillingssenteret*), a publicly funded organization (established in 1997), voiced a different opinion. It stated: 'According to the opinion of the Centre for Equality the legal father is a significant person in the lives of many children. In many ways, we understand the proposed legal amendment as a displacement of rights from the legal father to the biological father without having sufficiently taken into account the consequences for the child'. Their concern was whether knowledge of biogenetic origin per se is good, without due consideration regarding the disruption such knowledge may entail for the child, and stated: 'The intention of the law must work for the child's best interest and not harm it. [The Centre] does not see that the proposed changes ... will necessarily favour the child'.[17]

The question of rescinding the anonymity clause with regards to sperm donation was not phrased in terms of a father's right to ascertain his biological relationship to the child. In fact, such a proposition would undermine the very purpose of sperm donation as this is understood. Rather, it is the child's right to know its biological father that is stressed. This is seen to be in the best interest of the child. The legal introduction of named sperm is, as we have seen, based on an assumption that it is essential for a child to know its origins, and origins are understood as being the same as biogenetic origin.[18] In this vein, a person's genetic origin is perceived as tantamount to who he or she is. The strongest advocates of this position claim that this knowledge is fundamental; equivalent to a basic human right, it is unjust not to allow access to this information. Hence, sperm is attributed a defining quality: it not only creates identity, it is synonymous with it. It is also the biogenetic property of sperm that determines 'true' or 'real' (*virkelig*) paternity. In a certain sense, named sperm privileges the genitor–child relation, as it confirms the significance of shared substance (in kin relations.) With the use of named sperm, the child becomes visible in a different way and becomes perhaps a different child. We could say with Franklin that the future child embodies the law (Franklin 1999: 161): its very conception explicitly recognizes a double relatedness, that of both pater and genitor.[19] However, there is of course no doubt about who the legal father is.

The situation with respect to egg donation is quite different. Despite the fact that reproductive technologies also make eggs available, in the Norwegian context eggs cannot be made to move between women. Eggs – or egg cells – can only be made to move in and out of the same body. By insisting on the prohibition of egg donation, Norwegian legislators recognize one dimension of reproductive disruption – that is, the potential split between gestational, genetic and social motherhood – but they refuse to endorse it. In rejecting egg donation as a path to potential motherhood, the legislators deny women in need of an egg the legal means to obtain it. These women have to travel abroad to obtain one, and they also have to finance treatment themselves. Within the terms set by the law, motherhood has only one source; or, to put it the other way round, a child can only have one mother.

In order to understand what is at work in maintaining the distinction between egg and sperm donation, I turn now to the combined discourses, regarding these forms of exchange. It is the way sperm and egg donation are talked about – in the same breath as it were – that is revealing and gives us an inkling into the rationale behind their differential treatment.

Mater Semper Certa Est

I start by quoting at length the Ministry of Health's position, put forth in conjunction with the first legislation to regulate assisted conception, as these arguments echo and condense views held more generally. The Ministry states: 'There are those who argue that egg donation is in principle not different from sperm donation ... This is a view the Ministry rejects. Women's and men's reproductive functions are different – seen from both the donor's and the recipient's point of view'.[20] The Ministry continues:

> In contrast to sperm donation, egg donation requires medical surgery ... Donation of eggs has more similarities with transplantation than it has with sperm donation ... In contrast to sperm donation, egg donation creates a situation different from natural conception (*naturlig befruktning*).[21] Donor insemination does not break fundamentally with that which occurs in natural conception. Whether conception occurs artificially or naturally, the sperm is something that comes from the outside. This implies that there will always be some uncertainty as to who is the father of the child. With natural conception it is not unusual that there is a discrepancy between legal/social paternity and biological paternity ... Egg donation implies a fundamental break with that which occurs in natural conception. With natural conception the uterus and the egg constitute a natural unity. Conception, pregnancy and birth is a unified (*helhetlig*) process which occurs within the woman. With egg donation this unity is broken ... [posing] entirely new problems regarding the child's identity and roots. Until now, the borders/limits between the physical and social have been clear. With egg donation physical motherhood is split ... There is reason to believe that this lack of clarity will cause insecurity with regard to the identity of the child.[22]

A well known female Labour Party politician, Grethe Knudsen, following this train of thought during a Parliamentary debate on biotechnology, states:

> That women give birth has been so taken for granted that in Norway there are no rules of law defining who is the child's mother. Conception, pregnancy, birth have been a unified process. It is not that simple any more. And this separation – that conception occurs outside the body – implies that mother becomes more like father. Until now we have considered the mother and the unborn child as one. This has also been decisive for the right to self-determination in the question of abortion.[23]

In a later debate, the following statements[24] were made: 'With regard to the order of nature, egg donation is a significantly larger interference than sperm donation. Egg donation would be a breach of the inviolability and unity of pregnancy'. And: 'We do not know the consequences of introducing a notion of the strange/unknown mother (*fremmed mor*). The feeling of belonging to mother (*morstilhørighet*) is the most fundamental of human [emotions]'. Or in the words of one female politician: 'It is wrong that the mother of a child should be unknown. The belonging to mother (*morstilhørigheten*) is inviolable. For me this is not a question of gender equality, but respect for the order of nature. To permit sperm donation is less problematical than egg donation, first and foremost because sperm is easily available and cannot be regulated by law so that we can hinder a child from having an unknown father'.

In addition to the ideas of nature that are expressed with regard to natural procreation, there are at least two more points that can be drawn from these excerpts regarding the values attached to maternity. One has to do with the impossibility of imagining an anonymous birth, a child 'without a mother'. Yet in France, this is a practice which is protected by law: through *accouchement sous x*, a woman has the right to give birth anonymously.[25] This is something which would be unthinkable in Norway, where the unity between conception, pregnancy and birth is not only natural, it is inviolable. Furthermore, as is evident, the mother is given a unique status with respect to reproduction, and it is this status that determines what is right regarding egg donation. Recognizing the implications of reproductive technologies, Knudsen acknowledges that mother becomes more like father. Eggs, like sperm, are now available outside the body, and this shared quality has significant implications: maternity and paternity may become equally uncertain. Where egg donation is allowed and anonymous sperm donation upheld, this would certainly be the case. However, in Norway the opposite is occurring. Rather than rendering motherhood more like fatherhood, paternity is becoming more like maternity.

Disruptive Effects

With the advent of new kin categories – such as donor father, genetic father, social father, or co-father; or, with regards to motherhood, genetic mother, gestational mother, birth mother, surrogate mother, social mother and co-mother – both paternity and maternity have become ambiguous categories. The new situation is that motherhood has been disrupted in ways that fatherhood has not, and in some ways motherhood has taken on the traditional perceptions of fatherhood: it has become equally uncertain. The socio-cultural and legal responses to these new facts of life vary cross-culturally, and there are, as noted, disparities within Europe regarding egg donation. In the United States, issues pertaining to maternity rights have been taken to court – and courts have had to adjudicate competing claims to motherhood. The most famous of these, mentioned earlier, is the case of Baby M (see Dolgin 1997; Fox 1997). However, in Norway the situation is different.

Rather than accept these new facts as a condition of modern family formation and kinned relatedness, Norwegian legislators have opted for another path. In the face of these ambiguities about maternity and paternity, it seems that the need to establish some form of certainty is critical. Biogenetics serves this purpose well. In this case, certainty implies preserving the biological unity of motherhood as it is understood. It also implies standardizing biogenetic connectedness as a unique quality of each individual as this is perceived as a core element of identity. To paraphrase Cadoret in her discussion of different kinship truths and the role of genes, knowledge of genes (or birth) becomes the truth of a particular person; the part comes to stand for the whole (Cadoret 2009: 83–85). By appealing to the natural foundation of the unity of motherhood, biological relatedness is given primacy over other relations. Moreover, by tying identity to biogenetic origins and a need for certainty, these relations are also projected as fundamental. Therefore, it follows 'naturally' that the anonymity clause be repealed. (And also that new paternity laws, underscoring the importance of biological paternity, are passed.) Finally by evoking a child-rights discourse to ground the significance of knowing one's biological origin, nature, identity, rights and knowledge all work together to produce what is projected as an ethically correct position.

Collapsing notions of biology and rights, knowledge of biological origin is considered to be in the best interests of the child. This is claimed irrespective of the effect upon relationships that such knowledge might involve. This is particularly evident in the debates surrounding revisions to paternity laws. In this case the importance of establishing biological paternity coupled with a belief that it is in the best interest of the child to know its biological origins as early as possible overrides the psychological and emotional concerns of a stable and secure environment for the child – for example, in the case where a legal father with long attachment to the child is proven to not be the biological father. Thus, the certainty with which knowledge of biogenetic relatedness is endowed gains prominence over the emotional ties of social belonging. Nevertheless, the overall situation is not unequivocal. This is exemplified by the extensive practice of transnational adoption (see Howell 2006). This is not only a practice which many of the involuntary childless opt for but also one that is endorsed, and even preferred, by politicians and legislators. There is no doubt that there exists a public recognition of the fundamental significance of non-biological relatedness and that successful kinning of foreigners, in the sense of total incorporation into the bosom of the family, is possible. However, what appears to make adoption more acceptable (in contrast to the use of reproductive technologies) is that these children have been born naturally; they already exist, but are in need of a new family, home and care. Moreover, their right to know their biogenetic origin is already inscribed in law (more specifically, by the fact that Norway has ratified the UN Convention on the Rights of the Child).[26]

With respect to legislative processes, then, it appears that framing arguments with a child-rights discourse makes it easier to defend and accept government intervention. It is as if the notion of a child and its rights evokes an implicit yet shared understanding that things done in the best interests of the child are by definition good (and therefore right). Children are also subjects of the law, and the child needs someone to speak on its behalf. The state takes on this obligation through its legislation. In fact, in cases where no father has recognized a child at its birth, following the Children's Act of 1981, the state has the responsibility to determine who the father is.[27] As the mother of the child is the one who has given birth to the child, no such provision is necessary for determining maternity.

One implication of all these converging processes in contemporary Norway is that paternity appears to become more like maternity – equally certain, and not the other way round, as Knudsen envisioned. This, I suggest, may perhaps be one of the major shifts in procreative ideology, and the one most indicative of the importance granted biogenetic relatedness. Thus, the Norwegian case seems to substantiate Porqueres i Gené and Wilgaux's claim that there is a current trend to redefine father as *quem sanguis demonstrant* (as defined by blood) and can be interpreted as an attempt to make absent fathers present, or 'known', if only through establishing biogenetic connectedness (Porqueres i Gené and Wilgaux 2009: 122). The prohibition on egg donation and the abolishment of donor anonymity both underpin this idea of a perceived need for certainty. As socio-cultural phenomena they articulate a specific tendency to biologize identity and personhood in Norway. It is at this juncture that public discourse and legal actions cohere. Again, despite dissent, both egg and sperm are attributed unique qualities as sources of identity, and therefore it is best, in fact ethically correct, that the source (*opphav*) be 'known'.

Pater Vero? Turning the Tables

So far the argument is fairly neat – that is, as long as I remain within the parameters of legality and the debates surrounding legislative processes. In practice, things have a tendency to become more confused. I present a case that flies in the face of the law and gives the arguments about paternity – and maternity – yet another twist. The example is one that was not (until very recently; see Chapter 6) even contemplated by policy makers, yet is another effect of both the potentialities of reproductive technologies, of globalization and of the law. Its subject is a male homosexual couple, their desire for a child and how they realized this dream. The case was presented at a seminar I attended about reproductive tourism, organized jointly by the Norwegian Biotechnology Advisory Board and the National Medical Museum. The speaker was introduced as Odd Jenvin, 'father of two children born by a surrogate'. The title of his talk: 'How Does Norwegian Bureaucracy Meet the Children who Arrive by Cyber Stork?'[28]

Jenvin had travelled to the USA together with his partner in order to have children by a surrogate. He stressed the fact that he uses the term surrogate and not surrogate mother. This is because he and his partner chose for their children to grow up with two fathers. As he says: 'In our family there is no mother. And we do not want to give the impression that a mother exists somewhere. This does not mean that they will not have contact with the woman who gave birth, but it means that she is not their mother'. He goes on to say: 'Many politicians and experts are not used to this way of having children.

But that it is uncommon does not mean it is wrong ... [It seems likely] that more and more people will accept this way of having children as legitimate. I think that political processes and in due time legislative processes will be marked by this development' (Jenvin 2008: 32–33). According to Jenvin there were about 100 children in Norway at that time who had been conceived by a surrogate; most of those in need of a surrogate travel to the USA, where several states accept same-sex parenthood.[29] He and his partner went to California.

After presenting some details about the practice of surrogacy, Jenvin focused on how children born by a surrogate are entered and registered in Norway. In particular he was concerned about children born to male same-sex couples. These children (in contrast to those born by surrogate to heterosexual couples) run into particular problems, he claims. It appears that at the time (that is, 2007) Norwegian bureaucracy had problems with their categories: there was no way to register a child with two fathers, despite a birth certificate issued in the US certifying that the two men are the parents of the child. In some cases, these couples have been asked to take a DNA test in order to ascertain who the biological father is, so that the non-biological father can then proceed to apply for step-child adoption. In other cases, the child has been registered in the National Population Register, but without parents; however, they are given a social security number (*personnummer*), which is essential in Norway in order to be able to access all public services such as health care, kindergartens, schooling and so on. In other words, officially the tie between the parents and children has been severed. As Jenvin says, 'we then have a child of a few months living alone at an address where two men also live' (Jenvin 2008: 36). In a few cases, same-sex couples have been able to register a child born abroad without a problem. He also notes that single fathers (with a child born by a surrogate) do not seem to encounter the same problems as same-sex couples. He concludes that there seems to be a somewhat arbitrary practice and that this is related to a too narrow interpretation of the Norwegian child legislation that says that the mother of a child is the one who gives birth. 'From this can be deduced', he continues, 'that a child can only have one father and one mother. A child cannot have two fathers' (Jenvin 2008: 36).[30]

This case is illustrative in many ways, adding yet another dimension to the issues discussed above. In addition to demonstrating the inadequacy of the state bureaucracy's ability to confront reality in a meaningful way (for those involved) it implicitly raises questions of illegality and what is considered morally right. The case also, and importantly in this connection, questions the very notion of motherhood, and by implication fatherhood and parenthood, and what constitutes a proper family. Moreover, the example challenges the notion of infertility and what it is to be involuntarily childless (and hence who has the right to infertility treatment within the national health-care system).[31] In so far as infertility is based on some pathological definition, a homosexual fertile man cannot be considered infertile. There are many who would agree that it is a natural fact that men cannot have babies (and representatives of the

Norwegian Lutheran Church are especially articulate on this point, but they are not alone). The fact that a man is gay is therefore no argument to qualify for infertility treatment.

The Jenvin couple's actions pit notions of nurture against nature (heteronormativity and biological connectedness), where the act of caring (Borneman 2001) is privileged over the act of begetting. These actions also challenge the fundamental principles couched in the Norwegian Biotechnology Act, while demonstrating their effects. The Jenvin children are born of a surrogate with the use of donor eggs, the sperm provided by the two men. Thus, they 'break' the law on two counts. However, Jenvin and his partner are the biological parents of their children. Yet, as Jenvin observes, these practices are being introduced in Norwegian society; they are becoming public and they are producing effects. In the following I will limit my discussion to surrogacy.[32] Surrogacy sets a totally different agenda with regard to kinship relations, not only because it challenges established notions of motherhood (egg donation does too, as we have seen), but also because it allows for same-sex male parenthood that is not based on adoption. Moreover, it brings transnational reproductive practices back home.

It follows from what I have argued above that cases of surrogacy where donor eggs have been used are untenable on several counts, as such a practice defies Norwegian legal definitions of motherhood: neither genetically nor by birth is the child to be related to its intended mother. This also has repercussions for the legal status of the children. A recent and controversial surrogacy case in India confirms this point. It concerns a Norwegian woman, Kari Ann Volden, who had entered a surrogacy contract with a clinic in India. Using donor sperm and donor eggs (whose 'nationality' is not known), the Indian surrogate mother gave birth to twins. In order be able to bring the children to Norway, the children needed passports. Moreover, Volden's status as the legal mother had to be cleared. When the Norwegian authorities learned that the twins were born to a surrogate using donor eggs, Volden's request for passports was denied, as was her later application to adopt the twins. She was stranded in India for over a year. The significant point for rejecting her adoption application is that she had no genetic relationship with the children. Had Volden used her own eggs, the adoption procedure would have been authorised, as Volden would then be biogenetically related to the twins. This would also have resolved the issue of passports, as Norwegian citizenship is primarily based on the principle of *jus sanguinis*. However, Norwegian adoption legislation does not cover children commissioned through a clinic with the use of donor gametes. Thus, Volden's application was denied. Having no legal Norwegian mother, the children also had no right to Norwegian citizenship. Moreover, the Indian authorities would not recognize the children as Indian, hence the two children were *de facto* stateless. However, the commercial aspects of the case (the woman bought eggs and sperm and paid for a surrogate), and the fact that the transaction was done in India, have definitely compounded the issue, bringing to light the many ethical aspects of this kind of reproductive transaction. In April 2011, the situation was resolved, at least in so far as the Norwegian authorities took responsibility for the two children, who were allowed to travel to Norway. Once there they were given a guardian until the adoption process found a viable solution.[33]

However, if as I have suggested, and the above case illustrates, biogenetic connectedness is paramount, the use of a surrogate to gestate a baby with the intending mother's own eggs should not be problematic.[34] Yet, within the terms set by the law, this form of surrogacy still renders the category 'mother' equivocal as it undermines a notion of unitary motherhood (*helhetlig morskap*) and mother-belonging (*morstilhørighet*). Thus biogenetic relatedness 'alone' is not enough. In either case, the practice of surrogacy breaks fundamentally with Norwegian legal notions of motherhood, as 'the mother' is not evidently known.

Significantly, surrogacy (in the case of homosexual men) also breaks with notions of fatherhood as these have been perceived. Let me repeat what have been the common assumptions: paternity is 'uncertain' by nature; however, a child will always know who its mother is. Thus the idea of a 'fatherless' child is — or has been — accepted, but the unknown mother is not only unthinkable, it is wrong. Yet, as we have seen in the case of Jenvin and his partner (and their two children, born as twins, where each man is father to one of them), the fathers are known, but the mother is not. In fact, for them it is essential that the mother not be known as 'mother' but rather as the woman who gave birth. Thus Jenvin and his partner have effectively turned the tables: they deny the birth mother the status of mother (they also obfuscate the egg donor), privileging fatherhood as the unique status. In many ways, their procreative act can be seen as radical, even subversive. It is an act that on one level defies Norwegian values of relatedness and nature. On another level, however, their act feeds into existing notions of fatherhood.

Yet there is a problem. And yet again the problem has to do with establishing the 'true' father through biological connectedness. According to Jenvin, Norwegian authorities will not readily accept male same-sex parenthood. They (the authorities, that is) propose a simple 'biological solution', suggesting that the two men each take a DNA test in order to ascertain biological fatherhood, and then apply for permission to adopt the child of their (male) spouse – so called step-adoption. This solution is unacceptable for the Jenvin couple. Step-adoption undermines their mutual project of parenthood as it does not recognize that the child belongs equally to both of them, irrespective of biogenetic ties. The authorities will accept a single father (with an unknown mother) but cannot accept two 'unknown' fathers. And the question turns on what 'known' means. To the Jenvin couple, all their children need 'to know' is that they are the children of two fathers, born by a surrogate in the USA (who they also 'know'). Thus, while excluding a mother, they nevertheless confirm the traditional attitude to paternity: that it is uncertain. To the authorities, however, it is a matter of 'knowing' who the 'real' father is in order to ascribe proper filiation, and this can only be established by proving biogenetic connections. To my mind this is another twist on the drive for certainty that a biological principle propels.

At the heart of this matter is an idea of family and that it is unnatural that two men should procreate. Family values are indeed important in contemporary Norwegian society – so much so that for the involuntary childless the significance of being a family, having children, overrides many other concerns. Thus the desire of same-sex couples to be a family with children is congruent with values considered fundamental in Norwegian society.[35] However, the way they choose to become a family is problematic. Despite the revised Marriage Act of 2008 granting equal rights to same-sex couples, male homosexual procreation challenges the heterosexual norm and the privileging of motherhood. The following case (taken from Spilker 2008) is an apt illustration.

The former Christian Democratic/Conservative coalition government proposed a definition of the family which de facto includes everyone: 'the family comprises married couples with or without children, couples cohabiting with or without children, homosexual partners with or without children, single parents who live with children ... families with foster children and single people living alone' (quoted in Spilker 2008: 119). Yet as Spilker demonstrates, the same government proceeded to clarify that some families are more valid than others, thereby introducing a hierarchy of values. The ideal family that serves the best interests of the child is one based on a heterosexual norm and which by implication ensures that knowledge of biogenetic origin is known. Hence, the involuntary childless (including here same-sex couples) and the state agree on the significance of being a family. However, they disagree on what constitutes meaningful relatedness in family formation.

In the last instance, this is what the Jenvin case demonstrates: by insisting on being recognized as a family with two 'unknown' fathers, they are creating new forms of relatedness. Although the Jenvin couple concur with the state in so far as they reject multiple parenthood (a child should only have two parents), they are at the same time undermining central tenets in Norwegian understandings of kinship and filiation. Although their marriage is legal, their procreative practice is not. And although their parenthood is based on biogenetic relatedness, not only is this only partial (as the mother is not recognized), it is also in a sense uncertain (as they do not wish to specify who is the biological father of which child). They are not willing to accept knowledge of biological origin as the 'truth' about the identity of their children – and about their relationship to them. In other words, they deny that the biological truth of DNA (to establish paternity) is the relevant fact, and insist on their mutual parenthood. By refusing to take the DNA test, they also refuse to establish certain filiation, undermining the significance of biological principle. It is not altogether obvious whether or not this particular procreative practice also confirms the trend that fatherhood is becoming more like motherhood, although it does bring into existence a new 'father', potentially placing 'father-belonging' on par with 'mother-belonging' (*morstilhørigheten*). This is at least my reading of what is going on. What is obvious is that children born to this marriage will, like other children of many heterosexual marriages, not 'really' know who their father is (unless the couple complies with the request of the authorities and submits to a DNA test). However, and more significantly, they do not 'have' a mother. Thus, these children do not embody the law – rather, they fly in the face of existing legal categories.

By Way of Conclusion

This chapter has focussed on public negotiations of kinship and family relations as these are articulated and grounded in legislative processes. More specifically, I have been concerned with notions of maternity and paternity and the ascription of proper filiation. The consistent prohibition on egg donation, the rescinding of the anonymity clause regarding sperm donation and a case of surrogacy have served to highlight two main points (and their effects): On the one hand, a tendency to biologize identity in the sense that the biological principle is given priority in ascribing 'true' filiation; on the other hand, a drive for certainty, insisting on the significance of knowing your origin – *opphav*. These two factors are contingent and mutually implicated – one serves as an argument for the other. One effect, I have argued, is that in some senses paternity is becoming more like maternity: equally certain.

In the face of the new reproductive technologies and the knowledge they bring with them about the indeterminacy of biological facts, what is striking in the Norwegian case is the way biology is (re)produced as natural – for example, the idea of 'mother-belonging' and the fact that paternity is always uncertain – while at the same time the natural – or respect for nature – is given as the reason for restrictive regulations. Natural facts produce, it appears, natural values, and these facts and values are discursively conflated. However, and as the Jenvin case makes explicit, in practice this conflation creates problems – the facts are interpreted differently and given different values by the authorities and those who wish to make use of reproductive technologies. New forms of family formation effectively challenge the intentions of the law, raising questions about the limits of state intervention in the face of individual autonomy and procreative choice. The next chapter takes this question further. Again I address a problem of conception, but the issue is not that of having an 'own child' but rather whether a person or couple should be able to decide 'what kind of child' they are to have. In other words it is about selection – and hence about the 'sorting society'.

Chapter 5

THE SORTING SOCIETY:
KNOWLEDGE, SELECTION, ETHICS

> Today, in the domain of bioethics, what is taken to be at issue is the fate of humanity ... What is at issue is a crisis of 'dignity', the symbol enshrined in the Universal Declaration of Human Rights as the bulwark against the justification of any future Auschwitz.
> —Paul Rabinow, *French DNA*

> Is a fertilized egg a person? The question remains open.
> —*Aftenposten*, 24 August 2003

> The biopolitics of the Socialist Left Party can create a sorting society that is reminiscent of Hitler's Germany.
> —Inger Lise Hansen, Christian youth politician, *Vårt Land*, 1 July 2005[1]

> Bishop Ole D. Hagesæther warns the Church Council against permitting research and testing for illness on incipient human life.
> —*Vårt Land*, 21 June 2006

> The debate on early ultrasound. The public domain created a sorting society years ago.
> —*Dagsavisen*, 7 March 2011

Reproductive Choice

In 1994, the legislation concerning 'artificial procreation' came under a more encompassing Biotechnology Act that included the regulation of technologies that impinge on reproduction, such as preimplantation diagnosis (PGD) and prenatal diagnosis (PND).[2] Norwegian policy makers at the time – as in 1987 – acted so as to limit the application of reproductive technologies. This they have done on the basis of a precautionary principle – and an awareness of what scientific progress in conjunction with these technologies might bring about. Underpinning the decisions of the policy makers are images of a possible future society and a will to direct its course, whether or not they are agreed on what this course should be. A significant question that underlies many of these deliberations is: Who should have the right to decide on such vital matters as procreative practice? One intentional effect of the restrictive legislation is to curb individual choice. Yet, people choose to act in defiance of the law, doing what they feel is right for them. And in this global biomedical market, the possibilities are legion.

There is no doubt that biomedicine and choice operate together. The available biotechnologies lay open a field of alternatives catering to very different 'needs'. Moreover, the very commercialization of biomedicine and its insertion in consumer cultures implies choice. However, at issue are not only choice but also the way that choice is structured. In some countries – for example, the USA – the biomedical industry is organized around an idea of consumer choice, endorsed by a culture – or an ideology – that sees choice as central to its ethos. For example, Becker, in her discussion of reproductive technologies and gender in the USA, states: 'In the United States choice stands for autonomy, independence, and freedom of will, signifying women's sense of themselves as having an influence on the process in which they are engaged' (Becker 2000: 242–43). She suggests that the lack of regulation in the USA is unique, and goes on: 'Respect for individual autonomy is apparently a major factor in contributing to technological interventions. Because American ideals favor minimal intervention by government, regulation has been left primarily to the industry itself' (Becker 2000: 247). This is in stark contrast to Norway, and is precisely the situation that legislators wish to avoid. Here, as in other places of the world, individual choice in procreative matters is not a given. Quite the contrary, states – or religious authorities – may regulate to impede individual choice.

In welfare states such as Norway, which provide public health-care coverage, this implies that some procreative services (those which are not 'legalized') are not covered by the social security system. That is one sanction implied by the law. Nevertheless, the actions of the state may be viewed as merely symbolic (or even hypocritical)[3] in so far as people with the means can easily travel abroad and gain access to treatments not permitted in Norway. In fact, this very possibility has been used as an argument for more permissive regulation, on two grounds: to ensure that people obtain safe and secure treatments; and, perhaps more importantly, to ensure that people in Norway have equal access to treatment. In other words, to ensure that the law does not work against those who do not have the economic means to make use of private clinics, be they in Norway or abroad. Thus, politically, a principle of universality is voiced – through an equal-access argument – but which nevertheless has not won ground. Other ethical issues have been deemed more important.

So far, I have mainly focused on those reproductive technologies that have to do with assisted conception. My gaze has been directed at the different ways in which people come to terms with the challenges that assisted conception pose with a view to kinship, and especially filiation. The perspectives of both the involuntary childless themselves and policy makers have been explored. I have also detailed the process by which the present Biotechnology Act has come into being, pointing to some underlying trends regarding maternity and paternity. Keeping within my overall framework of reproduction, this chapter shifts tack. It is not kinship or relatedness that frames my analysis. The issue is not that of having an own child (at all) but rather a question of 'what kind of child'.

In what follows, I address quite another problem of conception. The subject matter is the embryo (and fetus) and the way the embryo is not only constituted but also constitutive of the social field surrounding it (Franklin 1999: 163). In addition to the meanings attached to the embryo, this field includes matters of choice and the possibility of selection. This chapter pursues this focus by delineating another contentious site, that of the 'sorting society' (*sorteringssamfunnet*). The notion of the sorting society, in addition to creating an ethical publicity, mobilizing an idea of equality, indicates the role knowledge plays in certain alignments between biotechnology, the state, the individual and society. Working at the interface between individual and society, between personal choice and collective good, the notion of the sorting society draws attention to a socio-cultural dilemma. I argue that the sorting society gains its persuasive power from an implicit evocation of eugenics and an appeal to ethics. Ethics constitutes a productive site where the relationship between law and biotechnology is played out.

In his book *The Politics of Life Itself*, Rose (2007) suggests that the vital politics of our own century is different from that of the early twentieth century: 'It is neither delimited by the poles of illness and health, nor focussed on eliminating pathology to protect the destiny of the nation. Rather it is concerned with our growing capacities to control, manage, engineer, reshape, and modulate the very vital capacities of human beings as living creatures' (Rose 2007: 3). Rose is concerned with potential futures – an emergent form of life – and draws on the past to indicate what it is that makes our time different.[4] One of the pasts he evokes and which is relevant for my further discussion in this chapter is that of the eugenics movement of the early twentieth century. With regard to current debates, Rose points to the way eugenics is invoked both to distinguish the present *from* the past and to link the present *with* the past (Rose 2007: 55). Eugenic strategies singled out reproduction as the major field of intervention; at issue was the question of procreation – and who should be able to procreate. Thus there is a common associative ground to be made with the current use and regulation of reproductive technologies. For Rose, however, eugenics represents a specific form of biological politics, qualitatively different from that which characterizes present biopolitics. He makes a point of distinguishing the biomedical body from the eugenic body (Rose 2007: 44), and argues that contemporary biomedical interventions should not be described in eugenic terms. The term 'eugenic' should be reserved for strategies used to regulate whole populations. Thus he argues that 'life may today, more than ever, be subject to judgements of value, but those judgements are not made by a state managing the population en masse' (Rose 2007: 58). Moreover, underpinning eugenic strategies was an ideology of differential human worth; in fact, it was this ideology that made eugenic practices possible.[5] This is in stark contrast to contemporary ideologies based on human and equal rights.[6]

Accepting Rose's argument, I nevertheless want to bring forth an example that may shed a somewhat different light on the current situation. My example has to do with the moral status of the embryo and the biomedical developments that involve genetics and the detection of an impaired fetus – that is, those technologies that enable selection: preimplantation diagnosis (PGD, allowing for embryo selection) and prenatal diagnosis (PND). Again I draw my empirical reference from Norwegian debates (in Parliament and elsewhere) on the use and application of such technologies. I want to illustrate how a specific 'politics of life' takes form and I also want to suggest that these politics succeed precisely because they act on the sentiments, beliefs and values of their political subjects – that is, by action on ethics (Rose 2007: 27). They do this by evoking an idea of a 'sorting society', a term, I argue, that implicitly alludes to eugenic practices. In projecting a desired future, this political rhetoric summons an undesired past. It is this discursive site – and the ethical publicity created in its wake – that is the focus of this chapter.

Simpson uses the term 'ethical publicity' in his exploration of the donation of body parts in Sri Lanka, stating that such publicity 'invariably draws on the core values of society, culture and religion to shape the motivation given in particular contexts' (Simpson 2004: 840).[7] The notion of ethical publicity draws attention to the public discourses within which ethics are couched while challenging us to grasp the motivational grounding of this publicity in varied socio-cultural contexts. In different ways, the idea of the sorting society captures and condenses a process whereby certain medical procedures and their accompanying technologies have become acute ethical issues and the focus of intense political debate.

The Sorting Society

The Norwegian notion of *sorteringssamfunnet*, which I translate literally as 'the sorting society', has no English or French equivalent, and to my knowledge it is not used in any of the other Scandinavian countries. The term suggests selection, discrimination and, more subtly, eugenics. What is striking about this discursive phenomenon is its persuasive power, both in terms of the imagery it evokes and in terms of the underlying ethics to which it appeals. Its connotations are negative: Norway should at all costs avoid being – or becoming – a sorting society.

Since the early 1980s, and with increasing intensity through the 1990s and into the 2000s, the notion of the sorting society has come to permeate certain discourses in the Norwegian public sphere. It was first used in the early 1980s in connection with discussions about medical genetic services and the possible extension of amniocentesis (Solberg 2003a; Kvande 2008).[8] The term then gained momentum in the 1990s, entering the public domain in conjunction with the nascent discourse surrounding bioethics.

The sorting society is a term which many of the actors engaged in debates about preimplantation diagnosis (PGD) and prenatal diagnosis (PND) are prone to use. These are practices associated with reproductive technologies, although they are not directed at infertility. Whereas PGD is a technique that allows for the genetic testing of a fertilized egg *ex utero* in order to identify a serious genetic disease, PND refers to techniques (performed on a pregnant woman) used to determine the condition of a fetus. Most common among these are ultrasound and amniocentesis. What both these diagnostic technologies share is their ability to identify abnormalities (be it in the fertilized egg or the fetus), and by implication the technologies entail the potential elimination of undesirable embryos or fetuses. Hence, the knowledge that the application of these techniques impart may in turn lead to a sorting out of individuals according to specific characteristics. It is the potentiality of this knowledge and how it may be (ab)used which is at the heart of the notion of the sorting society. Such knowledge entails the possibility of selective abortion. Selective abortion is not the same as selective breeding, which is most prominently associated with eugenics. Whereas the latter aims at regulating a population through procreative practice (that is, who is to be allowed to procreate) with a specific aim of bettering a population, the former eliminates the result of procreation; it is a sorting or selection according to specific criteria 'after the fact'. Both practices stand in stark contrast to notions of equality and both, I argue, evoke eugenics, albeit in different ways.[9]

The idea of a sorting society has slipped into Norway's vocabulary carrying with it an underlying morality. Moreover, it is used as if Norwegians know what it is and what it refers to. It is this quality of being self-evident that is striking. It seems as if the notion of the sorting society has had an almost immediate resonance, even though it is hard to document such a fact ethnographically. However, Google searches give an indication of the continued proliferation of this term: in March 2011 I obtained 35,700 hits, compared to 7,890 in December 2009, 1,506 in December 2007 and 535 in May 2005.

A Parliamentary Incident

In order to illustrate the work of the sorting society I start with a political event. The setting is a parliamentary debate held in conjunction with the passing of the revised Biotechnology Act in November 2003. At the time the government was a right of centre coalition, with a strong Christian Democratic Party that proposed the new law. Dagfinn Høybråten, the then Minister of Health, also had his say during the session. Høybråten was the leader of the Christian Democratic Party at the time and is what in Norway would be called a 'personal

Christian', representing the more conservative spectrum of the Norwegian Lutheran Church. His speech is an apt illustration of the way the idea of the sorting society is used to bring together science, politics and ethics, articulating views iterated by many others. The fact that he is a prominent politician gives some added weight to his words.[10]

Høybråten used most of his allotted time in Parliament to argue against the sorting society. He grounds his arguments in the controversies and public debates that biotechnology has provoked, not least regarding questions tied to notions of the human being; that is, to views about when human life begins, human reproduction, the moral status of the fetus and the fertilized egg, and questions related to the use of technologies in order to sort away (*sortere bort*) – or eliminate – undesirable genetic characteristics. Moreover, he framed his speech to Parliament with an explicit reference to ethics and ethical dilemmas, insisting that this law is ultimately about what kind of society we wish to live in. In that direction, he argues that it is paramount that research and development be controlled and regulated for the benefit of all, and not the other way round. He says:

'We cannot let the technologies rule us; we must take hold of the development and direct the technology in the direction we wish to follow'.[11]

In his speech to Parliament, Høybråten presents his own definition of the sorting society, stating:

> a sorting society is one that eliminates that which is not in line with what society perceives as "normal" or "desirable". With the help of the techniques of preimplantation diagnosis one can sort away fertilized eggs that do not have the desired genetic qualities. And through the development of better techniques for ultrasound examination and better and simpler methods for prenatal genetic diagnosis our knowledge of the fetus increases, something that may give room for the possibility of sorting those whom are "different". Such sorting is problematical. It implies that we will create a society that signals that there is not room for all … The fight against the sorting society is a fight for human dignity and for the right to be different … That we wish to fight against a sorting society is expressed in the preamble to the biotechnology law'.

In this passage Høybråten refers explicitly to the preamble to the 1994 Act.[12] If you recall, this preamble emphasizes that the medical applications of biotechnology are to be used for the benefit of everyone in a society 'where there is room for all' and in accordance with the principles of respect for human dignity, human rights and without discrimination on the basis of genetic constitution, and based on the ethical norms that form part of our Western cultural heritage.

Høybråten goes on to stress that this legislation is about the kind of society we wish to live in and that 'society has the right and the freedom to set the limits and framework for the practice of the experts in the field'. He insists that this debate is not about abortion and 'women's right to determine whether or not she will complete her pregnancy to term'; rather, it is about denying a woman 'the right to choose what kind of child she wants'. Høybråten expresses his satisfaction that there is consensus on the latter despite disagreement on the former. With these words, not only does Høybråten give specific content to the notion of the sorting society, but this notion is also promulgated from one of the most powerful rostrums in the country.

There are several points to be drawn from this speech. As the Minister of Health made evident: The fight against the sorting society arises in confrontation with the new medical technologies. In line with the preamble, it is framed as a fight for tolerance and the right to be different, and hence a fight for human dignity. The difference alluded to is inferred: it has to do with mentally and physically disabled persons. More significantly, his arguments are phrased in terms of society and not in terms of individual choice or rights. In this vein, selective abortion is framed as something qualitatively different from self-determined abortion; and along the same lines, prenatal diagnosis is defined as something different than routine ultrasound (which is offered to all pregnant women around week eighteen of their pregnancy).[13] In so far as the application of ultrasound can now reveal the same kind of information as prenatal diagnosis, the Minister argued that it should be regulated by the same legislation as PND. This was the heart of the matter. At issue is not the right to abortion on demand but rather whether the Norwegian state should endorse a policy that might lead to a sorting society – that is, a society which for instance allows the elimination of fetuses with Down syndrome detected by the use of ultrasound. Such elimination would be possible if diagnosis is carried out within the twelve week limit of self-determined abortion.[14] This would be tantamount to accepting that some lives are unworthy of living, thereby challenging the very notion of human dignity. It would also undermine a fundamental principle of equality in Norwegian society, potentially creating 'completely new classes of differences'. Such is the thrust of his argument.[15]

Several other representatives echoed Høybråten's views, both with regard to the relationship between society and technology and with regard to the dangers of creating a sorting society.[16] For example, Bent Høie, the Conservative Party representative, states:

> The Biotechnology Act is less about science then it is about the fundamental values we wish to have as the basis for the construction of our society ... We agree that we do not want a sorting society, that human dignity be graded, and that a human being shall not be created to be a tool for another human being. The challenge arises when politics meets science, when our goals are to be applied as concrete guidelines (*føringer*) for both science and medicine ... Society must by way of democratic processes put down some boundary markers to delimit the area within which medicine and science shall be permitted to operate ... [W]e are facilitating a development that *goes against* a sorting society and thereby a grading of human dignity (*gradering av menneskeverdet*).

Meanwhile, the representative (Gløtvold) from the Centre Party said: 'We must have as [a basic idea] that human dignity should not be violated by a differentiation through a sorting society'. Or, in the words of another representative (Woie Duesund) from the Christian Democrats: 'In spite of our different views on abortion, today's debate [shows] that we can stand together in our fight against a sorting society'.

Representatives from the Labour Party (who were against some of the proposed restrictions) chose a different tack. They were more concerned with choice, and not least women's self-determination. As Gunn Olsen said, with reference to the chapter on ultrasound:

> many have got the impression that this represents a greater limitation on a pregnant woman's self-determination, and therefore this chapter [of the proposition] is contested. The reason for this unease is that the majority of Norwegian women regard ultrasound examination as a routine practice during [their] pregnancy, [and do not see it as a means] to find abnormalities in the fetus ... I believe that most women would consider the term 'sorting society' farfetched in this connection. Women must have free access to obtaining knowledge about the fetus they are carrying ... We have confidence that women themselves can regulate this need.

Or, in the words of another Labour Party representative (Karita Bekkemellem Orheim): 'It seems as if fear of new technology overrules freedom of choice, pregnancy care and fetal surgery. But this tightening (*innstramming*) [of the law] also represents a lack of confidence [in the idea] that women can make responsible ethical choices. I feel ... that the government is hiding behind the fear of the sorting society (*skyver frykten ... foran seg*)'. And in the words of Kristoffersen:

> I do not believe that it is only the majority that considers that the individual human being has an autonomous value (*egenverdi*). A basic principle of patient care is the right to self-determination, [including] in that phase that concerns a woman's pregnancy ... I react when someone from this rostrum continually talks about the sorting society. I do not accept being evaluated on a scale, where someone claims that his or her position on restrictions regarding ultrasound should be ethically or morally superior.

The representative from the Progress Party (John Alvheim) stated that he sees the restrictions imposed as 'paternalistic (*formynderi*) as well as rendering suspect Norwegian research milieus and experts ... The government's prohibitions appear to me as if the government is more concerned with unborn life than it is with born life'.

Thus, the representatives from the Labour Party and the Progress Party voice their concerns for the individual's right to choose, for a greater trust in both patients and researchers, and in general for a more positive view of science and the significance of knowledge. Moreover, they do not accept being relegated to an ethically inferior position just because they have a different view on these vital matters – and are not afraid of a sorting society. Again we see how ethics are used to intervene in the space between politics and science and are mobilized to ground the morally correct position. The sorting society provides this link. It feeds into an already established ethos, that of the precautionary principle, and reflects the restrictive spirit that has underpinned the making of this legislative process. It also, and perhaps more problematically, undermines an ethic of individual choice and responsibility while at the same time appealing to an underlying value of equality.

As I have said earlier, on a general level the law articulates the fear of the legislators concerning the ends to which biotechnology can be used, thereby also reflecting a mistrust or lack of confidence in those developing and applying the technologies. The debates display a tension regarding attitudes to this new knowledge and the thrust of research, as well as a scepticism towards scientists and experts in the field, articulating one aspect of the relationship between law and biotechnology. This attitude has been succinctly captured by Jan Fridtjof Bernt, a recognized professor of law. In a newspaper article (addressing what he terms 'ethical policing' with regard to the protection of the fertilized egg) he takes issue with the following statement by the Norwegian government (issued in 2004): 'The precautionary principle has a specific significance in that it sets the necessary limits for medical research for future generations'. Bernt questions the relationship between individual and collective ethical dilemmas and the limits of legitimate state power – for example, to what extent can a political majority make use of the law and threats of sanction in order to gain acceptance for a controversial ethical position. However, Bernt's overriding concern is how (and by whom) knowledge (and research) about vital processes can be used. He concludes his argument by stating that: 'it is knowledge that the government fears', and that the government's position (at the time) is a 'modern version of the story of Adam and Eve: legislators must impede research that will give us knowledge that may tempt us to do unethical acts some time in the future' (Bernt 2004). His position is that research has no other immediate consequence then the knowledge that it brings us. This is in contrast to the opinions expressed by many of the legislators (see Chapter 3).

Bernt's statement is polemical, but his point is relevant. It brings to the fore how knowledge and state policies are intertwined with ethics. However, this fear of knowledge is not only a question of restricting research, it is also related to how individuals may act on certain knowledge. To return to the question of PND and the possible detection of an impaired fetus: It is the conjunction of PND and self-determined abortion that causes concern about the possibility of systematic selection; hence, a regulation is seen as necessary to prevent a woman (or couple) from being able to choose the kind of child she wants (and not whether she wants a child at all). However, implicit in this position appears to be an assumption that, left to their own devices, women will sort. Empirically this also seems to be the case. According to Berge Solberg (a Norwegian bioethicist), nine out of ten pregnant women who know they have a fetus with Down syndrome choose to abort, but very few have this knowledge.[17] Solberg suggests that the reluctance on the part of politicians to permit early screening does not necessarily indicate a devaluation of pregnant women – that they not be trusted with this information – but is rather an indication of the politicians' reluctance to be held responsible for the sorting that may occur if early screening is permitted (Solberg 2003b, personal communication).[18] Therefore access to such knowledge is withheld, at least by the public health-care services. However this is interpreted, the fact remains that, for some, there is a value placed on a certain ignorance regarding some specific conditions of the fetus. And it is the state, through legislation, that ensures that this be the case. Ignorance is in this case bliss, lest one unwittingly contributes to the creation of a sorting society. Yet, this ignorance is fictitious as women have access to private clinics that offer early ultrasound – and, moreover, many women choose to do so. Furthermore, the Red–Green coalition government of the present time (2011) has proposed amending the Biotechnology Act, offering pregnant women early ultrasound as part of routine pregnancy control. This proposal has spurred a new round of public debate.

Looking Back: Some Comments on Eugenics

It is beyond the scope of this book to discuss in any detail the policies and practices of eugenics in the early twentieth century. Nevertheless, some words regarding the Norwegian situation is pertinent, given that I claim that a notion of eugenics forms a significant backdrop to and resonates with 'the sorting society', from which the idea draws persuasive power, despite the fact that the term eugenics is not used in the Parliamentary debates. However, these debates turn on questions concerning systematic selection, a differential grading of human worth, questions of human dignity and the possible elimination of those who are 'different'. All these phenomena are associated with eugenics. Thus, my intention in the brief overview that follows below is to indicate that there is a certain continuity between the past and the present, and that practices associated with eugenics (especially sterilization laws and the occupation of Norway by Nazi Germany during the Second World War) have ramifications down to the present. That is, they form part of a relevant context. Another point I want to bring forth is the role of the Church in these processes. The Church – through its representatives – is also an agent, and Christian values (however they are articulated) also constitute part of the relevant context.

It is a fact that Norway (along with the other Nordic countries) had an active and positive attitude to eugenics or racial hygiene (which is the term mostly used) in the years between the First and Second World Wars (that is, before Norway was occupied by the Nazis in 1940). In their book on the development of genetics and biotechnology in Norway during the twentieth century, Nielsen, Monsen and Tennøe (2000) state that in the interwar years eugenics was positively associated with progress and science, optimism and modernity. However, in the years after the Second World War, allusions to eugenic effects were often used to discredit particular claims or positions. Eugenics had become synonymous with the misdeeds of the past, and in particular with the atrocities of the Nazis (Nielsen, Monsen and Tennøe 2000: 190). In the preface to the second edition of their book *Eugenics and the Welfare State*, Broberg and Roll-Hansen state that '[t]he spectre of Nazi eugenics hovers over today's public debates on genetic technologies in human reproduction. Since the 1970s the word eugenics has become strongly associated with Nazism. To characterize a practice or idea as eugenic has been to condemn it as totally unacceptable' (Broberg and Roll-Hansen 2005: x). Thus the years of the German occupation (1940 to 1945) have cast a long shadow over postwar social policies.

Nevertheless, both Nielsen, Monsen and Tennøe (2000) and Roll-Hansen (1999) are at pains to point out (and demonstrate empirically) that the eugenic practices that were instigated in Norway were not of the same order as the ones implemented in Nazi Germany. Most importantly, they claim, they were not racially motivated.[19] Eugenic practices in Scandinavia were promoted by liberal democracies and formed an integral part of social welfare programs (Roll-Hansen 1999: 202).[20] They were connected with the early efforts of establishing a welfare state (Broberg and Roll-Hansen 2005: 5) with the support of both liberals and social democrats. However, it is important to keep in mind that eugenics – and a keen interest in genetics and heredity – along with a concern for the quality of the population was part of the zeitgeist. Science was coupled to population policies, not only in Scandinavia but also elsewhere (Rose 2007: 54–65).

In Norway, it was first and foremost the sterilization laws introduced in 1934 that was motivated by a concern with eugenics.[21] According to Roll-Hansen, the decline in population growth, along with what was perceived as an increase in the number of mentally disabled, insane and other groups with low social capabilities, motivated these laws. The arguments were varied, ranging from humane reasons to economic and efficiency ones. The aim, however, was constant: to limit the procreation of 'inferior' social groups by passing a law that provided a legal framework for voluntary sterilization (Roll-Hansen 2005: 154, 171). And the problem, of course, is how voluntary the practice actually was.[22] Sterilization on medical grounds already existed, and the issue was to provide the legal means for sterilization using social or eugenic indicators. Eugenics is not mentioned in the legal text, but it seems clear that eugenics was an important motive for passing the law (Roll-Hansen 2005: 171). The law was passed with only one vote against.

The sterilization laws must necessarily be understood within a broader socio-cultural and political framework. Other significant issues pertaining to the relation between sexuality and society were also being discussed at the time. Parallel to debates on sterilization, questions of contraception and abortion were raised and gradually gained in acceptance within Norwegian society.[23] Birth control, family planning and the creation of clinics (*mødrehygiene kontorer*) offering such services to women were part of the sexual politics of the time. What these issues share is their convergence around the morally contested separation of procreation and sexuality (Nielsen, Monsen and Tennøe 2000: 89). For some, such as members of the Church, both contraception and sterilization were subsumed under the overall concern for the potential liberalization of abortion (Nielsen, Monsen and Tennøe 2000: 93). Thus, the Church accepted, albeit reluctantly, the legalization of sterilization.

There are many different factors that impinge on the eventual passing of the sterilization laws and I cannot do justice to the complexity of the issue here.[24] There are, however, a few points I want to draw forth that are relevant for my further discussion of the sorting society. One is the role of religion and the position of the Norwegian Lutheran Church; another is the relation between science and policy making. The introduction of sterilization laws in Europe in the 1930s was a phenomenon of Protestant countries, where the Lutheran denomination was dominant (Roll-Hansen 2005: 187), and was in marked contrast to predominantly Catholic Southern Europe. The Catholic Church was against any state intervention in reproduction, and in 1930 the Pope condemned sterilization, as well as abortion and family planning in general (Nielsen, Monsen and Tennøe 2000: 118). Both Nielsen, Monsen and Tennøe and Roll-Hansen point to the fact that in Norway there was little public criticism of sterilization from within the Church. Roll-Hansen uses the arguments of Pastor Ingvald B. Carlsen as an illustration. According to Roll-Hansen, Carlsen claimed:

> the law signified a radical break with earlier Christian norms for social work, [as] Christian ethics had seen the miserable individuals ... in society as the objects of Christian spiritual care and mercy. Now it was demanded that one should help prevent such individuals from being born. The traditional objection of Christians was that life is sacred – man should not interfere with the work of God. Still, Carlsen found the proposed law acceptable considering how much suffering it would help prevent. To interfere in favor of existing life against life which did not yet exist must be permissible. (Roll-Hansen 2005: 174–75)

In his discussion, Roll-Hansen is especially interested in the relation between science and policy, and between scientists and politics. He argues that 'scientific experts were prominent participants both in the public debate that set the political agenda and in the further decision making process'. He draws attention to the fact that many of the experts were critical of the way genetic science was applied and that their opinions were in many instances at odds with the 'strongly hereditarian views among the general public on mental retardation, mental illness and various kinds of social misbehaviour ... taken to be the result and therefore the responsibility of science' (Roll-Hansen 2005: 183–84; cf. Lavik 1998: 11–12). Roll-Hansen points to a discrepancy between lay and expert opinions, and to an active relationship between scientific expertise and policy making.[25] This relationship undoubtedly shifted ground in the aftermath of the atrocities of the Second World War, and as we have seen, takes on quite a different character in the legislation of new reproductive technologies, where many of the arguments are directed at the need and the obligation to control – or be ahead – of scientific developments. It is therefore not coincidental that one of the arguments against legislating AID in the 1950s was that such practice might contribute to selective breeding.

The close of the Second World War saw a shift in values, leading to a radical reflection on the ethics of science. Both the Nuremberg Code (1947), produced as a result of the Nuremberg trials of Nazi war criminals, and the Declaration of Helsinki (1964) were significant milestones in establishing the rights of the individual and the principle of informed consent for all experimentation on human beings (Roll-Hansen 1999: 207). Nevertheless, the Norwegian sterilization laws passed in 1934 were upheld until 1977, with only slight modification in 1961. (Most of the sterilizations carried out after 1950 were granted on the basis of social-medical indicators, family planning being an important motivation.)[26] In 1977 sterilization was removed from public control and became a private matter, between a person and their medical doctor. However, eugenic arguments were relevant in the 1950s and 1960s in relation to abortion debates (Nielsen, Monsen and Tennøe 2000: 195–97; Giæver 2005). In fact, the term 'eugenic indication' was used to designate certain conditions for which abortion was permitted (Syse 1993: 25).[27]

The sterilization laws were also implicated in the government's assimilation policies towards minorities, and these also involved the Church. In 2000, Per Haave delivered a report commissioned by the Norwegian government on the sterilization of travelling people between 1934 and 1977 (Haave 2000).[28] This report capped a process that had gained momentum in the 1990s. The overall issue was the injustice and abuse (*overgrep*) committed by the Norwegian government with respect to Romany and travelling people in Norway (and consequently their demand for compensation for the wrongs committed). More specifically, the question turned on forced sterilization. In 1998 the Norwegian government issued a formal apology for its violation (*overgrep*) against the Romany people, an apology that was reiterated in a White Paper.[29] Among other things, the White Paper points to the work of the Norwegian Mission for the Homeless (which operated with support from the Ministry of Social Affairs).[30] This Christian organization was especially zealous amongst the Romany and travelling people.[31] With the blessing of Church and state authorities, the Mission actively promoted and enforced sterilization.[32] In 1998 the annual meeting of the Norwegian Lutheran Church (*Kirkemøtet*) discussed the violence against Romany people. However, the resulting statement was rejected by the Romany peoples' organization, as it did not actively recognize the complicity of the Church. Only in 2000 did the *Kirkemøtet* give an unequivocal apology and ask for forgiveness for the role of the Church in the injustice and abuses perpetrated against the Romany people.[33]

This process of reconciliation is obviously significant. It not only acknowledges the wrongs committed by the Church and state authorities but also illustrates the political efforts made to contribute towards creating a more just society. This overriding concern is also echoed in debates about the sorting society. The part played by the state and the Church regarding sterilization practices, and their efforts at coming to terms with a discredited past, is relevant to my argument. Although there is no explicit link between this process and that of policies regarding reproductive technologies, the fact that the practice of sterilization initiated in the 1930s has reverberated down to the present suggests that there is at least a certain public awareness of eugenics and eugenic practices. And, more significantly, there is an explicit repudiation of what these practices entail.[34] To what extent the recognition of this specific past is confounded with Nazism more generally is hard to ascertain, but what remains manifest is the complicity of the authorities, both Church and state, in eugenics policy, and, significantly, an explicit rejection of the same, indicating a fundamental redirection of policy values. Thus I suggest that it is not unreasonable to assume that this awareness has filtered into the contemporary Norwegian public sphere (*offentlighet*). It forms part of a cultural subtext that nurtures the notion of a sorting society.

The Law and the Church

As mentioned in Chapter 1, various scholars have argued that Lutheranism is a contributing factor to the construction of the welfare state. Broberg sustains this position. For him, the fact that the Church in Scandinavia is a state establishment is of vital importance for understanding Scandinavian history, and 'it is important to note that consensus and cooperation are paramount in medical affairs as well as other areas' (Broberg 2005: 2). Significant parts of the Lutheran community in Norway have seen influencing the outcome of the legislative process related to reproductive technologies as essential. The Christian obligation to influence legislation is reiterated in several Church documents and the Church Council of 1989 states explicitly: 'it is the obligation of the Church to deliver the ethical premises that ground legislation. The principle of the protection of the fetus (*fosterets rettsvern*) is one such inviolable premise'.[35] The main concerns of the Church (with regard to biotechnology) are directed at the notion of the human being, the notion of human dignity, the protection of unborn life, and the institution of heterosexual marriage. These concerns translate into debates about the moral status of the embryo; meanings of conception; implications of PND and PGD (abortion, selective breeding); the practice of anonymous sperm donation (AID); and the rights of homosexuals. Within the Norwegian Lutheran Church, one central point of contention is whether complete human dignity applies at the moment of conception or whether it is a question of gradual humanity. Discussions also revolve around whether there is a specific Christian ethics that must be applied to biotechnology (with reference to the Bible) or whether Christian ethics are subsumed under what is called general ethics (*almen etikk*), where reference is to society as a collective and human rights more generally.[36] More recently some of these issues have been discussed at seminars organized on the topic of 'religion and bioethics' (with a focus on Muslim, Jewish and Christian ethics and practices) or through major research projects on the moral status of the embryo (Østnor 2008).[37]

The two pivotal technologies in the debate on the sorting society – PND and PGD – represent a major challenge for members of the Church because they address the distinction between human life and human being; and because they address the fundamental question of human worth and human dignity. Whereas PGD is tied to meanings of conception and further reflected in meanings attributed to the fertilized egg and fetus, PND evokes the fundamental principle of the equality of all human beings, irrespective of their capabilities. Both, however, permit sorting, and hence some form of selection. One problematic question is whether the fertilized egg and the fetus are essentially the same.[38] There is no consensus among Christians on this point: For some they are different entities, for others they are one and the same.[39] Regarding the latter, the argument is grounded in the idea that complete human dignity begins at the moment of conception. However, one quality the two positions do share – regarding the fertilized egg, the embryo and a more developed fetus – is that they cannot speak on behalf of themselves. The Church sees it as one of its main obligations to society to speak on behalf of and protect the most defenceless. For some, the fertilized egg represents the most vulnerable of human life (whether it is considered to have complete human dignity or not).

The regulation of PND and PGD work at these troubling interfaces, even though the technologies imply different practices, are directed at different conditions and are applied at very different stages after conception. In addition, as we have seen, a forceful argument with regard to the sorting society has been the idea of equal human worth, the upholding of the notion of human dignity and the right to be different. Although it is difficult to identify these more general arguments as specifically Christian, I nevertheless suggest that the legislation could (or even should) be seen as a political transcription of some values, also held as fundamentally Christian. These latter, however, have to be inferred. Thus, what draws my attention is the conflation of secular values and more specific Christian values under the overarching premises of the preamble, expressed as the 'ethical norms that form part of our Western cultural heritage'.[40] If you recall, one of the points of contention regarding the preamble was precisely whether it should read 'Christian cultural heritage' rather then 'Western'. 'Western' gained the majority vote, subsuming Christian heritage. The preamble also constitutes the frame of reference for arguments pertaining to the sorting society, both for and against.

Ethical Dilemmas/Ethical Publicity

Keeping in mind that there is no consensus in Norwegian society on the use of these technologies, and that Norway had a Christian Democratic-led coalition government which was at pains to pass a restrictive biotechnology law, it was paramount for the government to find a way to achieve some form of consensus – that is, to find ways of addressing the most controversial issues in terms that would embrace society. The notion of the sorting society fits this purpose well.

I believe that the persuasive power of the sorting society is that it appeals to an idea of society to which Norwegians do adhere, while simultaneously glossing over some of the major controversies and inconsistencies that these technologies entail. Moreover, it does so on the basis of a general ethics, compatible with a secular society. It is not pitched as a debate between science and religion or faith against reason (see Mulkay 1997); rather, it is framed in terms of individual and society. The sorting society is used to summon universal values that are deemed central to Norwegian society – equality, respect, human dignity and human rights – and it does this by pointing to the dangers of systematic selection and a grading of human worth, thereby implicitly evoking images of racial hygiene and eugenics, reminiscent of a social order which is associated with the atrocities of the Nazis. These universal values are not framed as specifically Christian, but they are concordant with, as well as promoted as pertaining to, a Norwegian Christian world-view. At least that is my reading.

As I have indicated, one of the main controversies that the sorting society encapsulates is tied to the possible selection of embryos or fetuses that PND and PGD give rise to. The application of these technologies reveals information about the embryo or fetus. The mere existence of such potential information provokes a need – or even right – to know, and from this knowledge follows an obligation to act (cf. Strathern 1995). Thus it is the consequences, the implications, of knowing – whether an embryo or fetus is in some way impaired – that somehow underpins this debate. The possibility of knowing activates a distinction between abortion on demand and selective abortion. This distinction, I argue, discloses a tension between individual and society. Selective abortion is framed in terms of society by evoking the kind of society Norwegians ought to aspire to, as a kind of collective good. Self-determined abortion is framed in terms of individual rights and choices. However, this distinction grows out of a particular historical context. The former arises out of the potentials of biotechnology and has as its main site of contention the status of the embryo; the latter is the result of a feminist struggle for self-determination, with the woman's body as the main point of reference.[41] The effects of the distinction are a cultural – and ethical – dilemma.

One reading of this dilemma is the following: how to uphold women's right to self-determination (abortion on demand) while at the same time denying her the possibility of choosing to abort a seriously impaired fetus (or not implant a defective fertilized egg)? The only way to do this is to deny her knowledge of the condition of the fetus before the twelfth week. To deny a woman this knowledge, as I have suggested, is at the same time to imply that she is not capable of managing that information in an ethically correct way. (And I am now arguing within the parameters of the legal discourse, leaving aside the fact that women may act differently.) Another reading of this dilemma implies shifting the perspective from the individual to society. The problem then is not so much that women might choose to abort a seriously impaired fetus (for whatever reason) but rather that society – or rather the policy makers that make laws on behalf of society – cannot, or should not, be responsible for the sorting away of some particular kinds of individuals.[42] That would be tantamount to accepting a sorting society. It is at this point that a fear of systematic selection and eugenics implicitly makes itself felt.

It is also at this juncture that the notion of the sorting society does its work. It shifts the premises of the discourse from the individual (and her right to choose the kind of child she wants) to society while at the same time appealing to commonly held values of the welfare state: equality for all, tolerance of difference, human dignity and so on. It is a powerful argument as its appeals to a self-created moral community are as much about the past as they are about an imagined future. The notion of *sorteringssamfunnet* thus creates an ethical publicity about a society where there is room for all – while at the same time implying that to think or act otherwise is morally wrong. It forcefully signals that a certain use of these technologies is unethical, thereby undermining voices that might hold a different opinion. Significantly, within this discourse the cost factor is silenced; it is a non-issue. It is deemed totally unethical to argue for selective abortion on economic grounds – that is, that the care of severely disabled people is expensive and a burden on tax payers.[43]

The notion of the sorting society nurtures a fear that if we do accept that kind of society we are not only allowing for the selection of desirable children but also travelling towards a society with less tolerance for physical difference, for disabled persons, for weakness. We are potentially travelling towards a society that violates human dignity. Thus, the sorting society is primarily about society's attitude towards disabled persons and their families – and not about the fetus or fertilized egg as such. The more fundamental Christian meanings attached to these – for example, the idea that full human dignity applies at the moment of conception – have been subsumed under a more embracing discourse. The rhetoric of the sorting society feeds on one of the most fundamental values of Norwegian society, that of the tension and multivocality inherent in the notion of equality (Lien and Melhuus 2009). That is perhaps the secret of its success.

The way the sorting society operates, it is limited to practices that have to do with biotechnology and hence overlooks all the other forms of sorting that goes on. That is one of the paradoxes of the sorting society. Obviously, there are many other systems of discrimination in Norway, and people are sorted according to many different criteria. Nevertheless, these forms of discrimination have not produced a specific rhetoric, nor an ethical publicity as effective as that of the sorting society.

The accomplishment of the sorting society has to do with the ability of this notion to transcend the individual, concealing some fundamental dilemmas, and to evoke a vision of a society that many Norwegians cannot possibly reject. Although the idea of the sorting society is in some sense vague, it gains its added value by appealing to abstract notions that are considered good. Working at the interface between scientific and medical expertise and state policies, the notion of the sorting society places certain aspects of biopolitics over and above the politics of everyday life and establishes an ethical – or as some would argue, *the* ethical – position. That, it seems to me, is the basis for the resonance of this notion in contemporary Norwegian society. Moreover, it is this resonance that gives leeway for governing, activating and legitimating a certain state control in matters of reproduction.

Some Final Remarks

Earlier I referred to Nicolas Rose and his discussion of eugenic politics. I will also end on this note. Rose states that, 'the experience of eugenics has left an indelible mark on the politics of individual and collective health, throwing suspicion on any form of state management of vital forces that do not operate under the signs of autonomy, consent, and individual rights' (Rose 2007: 70). As I have argued, the experience of eugenics has left an indelible mark – also in Norway. So much so that this very experience can successfully be evoked through the rhetoric of the sorting society. Yet, as I have tried to show, and in contrast to Rose, state management (in the form of legal regulations) can be mobilized precisely to avoid what is seen (by some) as reminiscent of eugenic practices, and in doing so privileges society over individual autonomy. By throwing suspicion on the potential abuse of particular technologies – such as those that might lead to the systematic selection of desirable offspring – the Norwegian government at the time gained support for its restrictive legislation. It did so in the name of human rights and a humane society, but at the expense of individual choice. To avoid the risk that some lives may be seen to have less value than others, and hence be sorted away, individual autonomy and choice have (in these matters) been forfeited for what is seen as a higher good: to avoid becoming a sorting society. However, significant voices have been raised defending women's access to prenatal diagnosis, even before the twelfth week of pregnancy, and as intimated, it is likely that in the near future, with the shift in government, the legislation may again be revised.

What is perhaps striking about this debate is how it avoids confronting the moral status of the embryo – that is, the fundamental question of whether to confer complete humanity at the moment of conception. If you recall, the issue was framed (by the Christian Democratic Minister of Health) as one that had to do with selective abortion and not self-determined abortion. To establish this distinction was, I believe, paramount. Had the issue been framed solely in terms of the moral status of the embryo, it would have been impossible to bracket the question of self-determined abortion. And had the question of self-determined abortion also become an issue, it would most likely have been difficult to gain the necessary support for the restrictive legislation.[44] By evoking the sorting society and the potential risk of the systematic selection of desired offspring, it was possible to sidestep the question of the moral status of the embryo altogether. Nevertheless, this issue is in no way resolved, despite the fact that the most recent revision of the Biotechnology Act has opened up research on supernumerary embryos.[45]

Before concluding I want to make one final comment. It has to do with knowledge and the values of knowing – or not knowing. As we have seen in the case of anonymous sperm donation (Chapter 3), an important argument for rescinding the anonymity clause was that knowledge of biological origin is seen as constitutive of identity. Hence, it is a child's right to know the identity of the sperm donor. The changing of the paternity laws (granting men unilateral access to ascertaining their paternity) also underscores the significance of biological connectedness to establish the 'true' father (Chapter 4). The Jenvin case again exemplifies this need to know (on the part of the authorities): by insisting that each father take a DNA test to establish 'real' paternity and then proceed to adopting their spouse's child, the authorities sought to ascribe proper filiation. All these cases have to do with information that concerns relatedness, in other words kinship knowledge (see also Finkler 2001). The value of knowing is viewed as positive. Preimplantation diagnosis (PGD) and prenatal diagnosis (PND) can reveal information about the condition of the embryo or fetus. Except for situations of serious hereditary disease (which would be 'kinship knowledge'), knowledge of such information is seen as potentially threatening the order of society and therefore is only to be disclosed in specific cases. In this case, it is the value of not knowing which is deemed right. Reproductive choice takes on new meanings.

Strathern has argued that some kinds of information come with built-in effects (Strathern 1999: 64–88). She is specifically concerned with kinship, 'how knowledge about how persons are related to one another is acquired from … information of biological process' (Strathern 1999: 65). She sees such information as constitutive. I would argue that this also applies to the information obtained through PGD and PND. However, the built-in effects are of a different order. In the case of biological information (such as DNA) establishing kinship relations, the effect is the certainty – the verification if you will – of a relationship that is important. This knowledge is personal, attributed at an individual level – with concrete relational ramifications. However, in the case of PGD or PND, the information concerns the condition of the embryo or fetus. It is deemed necessary to withhold this information because of its perceived societal effects. 'Knowing' the condition of an embryo or fetus may imply the systematic elimination of undesirable offspring. This is the effect that must be avoided and which is 'built-into' the information. Therefore 'not knowing' is proposed (and accepted) as the legal regulation (irrespective of the fact that women may want to know and act accordingly). Contained in the technology is a vision of a possible future society. Thus, in contrast to the more immediate and concrete effects of kinship information, the information that PGD or PND brings mobilizes a more abstract value, a more embracing relationality, namely that of society and a tolerance for difference. In that sense, the embryo can, as Franklin suggests, be seen 'as possessing a kind of kinship status not only to specific persons but to humanity in general' (Franklin 1995: 336). Moreover, it might well be that the refusal to pass on such information also avoids creating situations of choice – and decision making – which are difficult to handle. The value of not knowing can also be understood in these terms.

What is evident is that some kinds of biological information are imbued with a certain socio-cultural potency, with imagination even. It implies action, be it at an individual level (a decision to abort or not) or at the level of society (a decision on how or if to regulate). What is significant in this context, is not only – or even – that the Norwegian state has the power to regulate access to such information, and hence the knowledge it implies (although as we know to a certain extent this is illusory), but also that the state – through its legislators – may find it appropriate to do so, specifying the conditions and terms under which such information may be obtained. The working relations between biotechnology and law could not be more explicit.

This in turn reflects, perhaps, a more general tendency that I have broached several times throughout the book that has to do with scientific developments within biotechnologies and the interface 'where politics meets science'. It is not as if Norway – neither the Norwegian public nor the politicians – is inherently sceptical to new developments in science in general. Quite the contrary. However, there is no doubt that reproductive medicine and biotechnologies are – and have been – especially worrying, and hence contested. The discourse about the sorting society reiterates the felt need to control developments in medical science, which were also evidenced in the debates about assisted conception. A regulation is brought into being to avoid practices that science makes possible. Thus it seems difficult to accept Bernt's claim that research ('science') has no other immediate consequences than the knowledge it brings, as it is precisely the potentiality of this knowledge (its built-in effects) that legislators (at least some of them) wish to contain. Yet, just as technology is socially informed '[where] specifically desired ends are built into the knowledge and techniques associated with biomedicine, the biosciences, and biotechnology' (Franklin and Lock 2003: 5), so also is legislation. Scientific progress is in a sense embodied in the very laws that the Norwegian legislators propose, as biological facts which are incorporated into the legal discourse.

Chapter 6

CONCLUDING REFLECTIONS: LEGAL (UN)CERTAINTIES

> Modern law is a continuous attempt at fixity and closure which is undermined by the impossibility of its own project. As a system of codification, whether based on legislation or on case law, it is an abstraction, and in consequence actual life-situations fit into it only imperfectly. This creates a zone of moral or at least pragmatic ambiguity and a room for manoeuvre … and where decisions can be made on strictly legal grounds which fly in the face of commons sense or everyday morality.
> —Olivia Harris, 'Inside and Outside the Law'

Global Reach, Local Appropriation

In Norway, as elsewhere in the world, new reproductive technologies have had an ambivalent reception. These technologies have made possible forms of conception that, separating procreation from sexual intercourse, permits the fertilization of eggs and the creation of embryos outside the body. Thus what has been perceived as the natural link between conception, gestation and birth has been severed. An embryo can potentially be inserted into any woman's uterus – irrespective of who provided the egg. This possibility has given rise to surrogacy, a practice that only over the past few years has become more common in Norway – and hence also a more urgent issue for policy makers. The cryopreservation of gametes and embryos permits the postponement of procreation to a later date, even posthumously – as, for example, in the case of Diane Blood (see Simpson 2001). Moreover, linear descent can be reversed by using grandparents as surrogates and children as gamete donors (Thompson 2005: 12). It has as, Dolgin notes, 'become possible to manipulate the spatial and temporal dimensions of reproduction' (Dolgin 1997: 2). Moreover, the embryo may be examined for genetic disease and discarded if found to be defective, while advances in prenatal diagnosis allows for the examination of the fetus and its possible elimination. The human embryo is also the source of pluripotent stem cells, making research on embryos a promissory field. This range of possibilities creates hopes for some, and profound unease for others.

Whatever the reaction, there is no doubt that these technologies are implicated in contemporary processes of social change, at the level of both individuals and society.

Reproductive technologies are a global phenomenon, but ones which, as all technology, is always localized. Their local appropriation and incorporation vary not only across the world, but also within regions of the world. These variations have different import for different social agents. For scholars interested in socio-cultural systems of reproduction and procreation, the various responses to assisted conception and associated technologies represent an exceptionally fruitful point of entry into exploring more encompassing socio-cultural processes within a cross-cultural perspective. Working at the interfaces between kinship, science, religion and politics, reproductive technologies are revealing of fundamental values. These values are often articulated with reference to nature or what is natural – that is, those areas that appear as self evident, or what we might call cultural blind spots. Thus, an examination of these values draws forth some of the implicit assumptions that contribute toward the constitution of society.

However, the global reach of assisted reproductive technologies (ARTs) is also important for those wishing to make use of them, be it the involuntary childless, medical personnel, infertility clinics or investors. Moreover, biomedicine is a promising area of research, and scientists all over the world are engaged in refining techniques as they develop new technologies. The mere fact that different political regimes have different regulations has made reproductive options a transnational phenomenon. What cannot be done at home can often be done elsewhere. Reproductive tourism has become a lucrative industry; infertile persons or couples are active consumers. The increase in cross-border reproductive practices has become a significant bioethical concern, especially with regard to the use of 'the reproductive capacities of poor women in transitional economies to complement the reproductive deficiencies of women in more affluent ones' (Donchin 2010: 323). Addressing this question, Humbyrd argues that 'the only valid objection to international surrogacy is that surrogate mothers may be exploited by being given too little compensation', suggesting that '[i]nternational surrogacy is ethical provided it is practiced following the principles of Fair Trade' (Humbyrd 2009: 112). Thus yet another dimension – that of 'fair trade' – is introduced into the frame. With reproductive technologies, problems of conception have become a significant dimension of biopolitics, as they also are of much bioethics. This procreative universe embraces the two central dimensions of reproduction: the regeneration of life and the regeneration of society. My interest has been to examine the relations mobilized to push these regenerative processes in particular directions. As we have seen these are multiple and complex.

I have traced the reception of reproductive technologies in Norway through a legislative process. These legislative acts are one way of coming to grips with

ARTs, illustrating a particular localizing process. I have focussed on the making of laws – a process that began in the early 1950s with the first attempt to regulate artificial insemination by donor (AID) and that continues down to the present. The last revision of Norway's Biotechnology Act was in 2007; at the present time (2011) a new revision has been announced and is underway. (One of the contested proposals is to provide early ultrasound to all pregnant women.) During this period, the terminology used to describe the procreative practices associated with ARTs has changed: at its inception, these practices were spoken of as 'artificial insemination', 'artificial procreation' or 'artificial conception', with the ensuing child termed a 'test-tube baby' or 'IVF baby'. Today, the term 'artificial' is no longer used; it has been substituted by 'assisted' (as in 'assisted conception', a term I have also accepted and employ). And the term test-tube baby has become obsolete. This change in terminology is not coincidental. In part, it may even be intentional. Be that as it may, the terminology used shifts the perspective, setting a different biomedical agenda.

'Artificial' connotes an intervention in nature, a practice that is not 'natural'. Assisted, however, has quite a different ring: it is a question of 'helping' nature. This shift reflects both a process of naturalization that has occurred with regard to the application of these technologies, and a gradual evolvement regarding attitudes toward the natural and natural procreation – and the ideas embedded in what some (of those cited and I have interviewed) have expressed as a 'respect for nature'. Whatever the case, there is no doubt that however contested artificial/assisted conception has been, there is now in Norway an acceptance of and a certain perceived legitimacy for many of these practices. It may be as Franklin suggests:

> The necessity for technological assistance thus comes to be seen as a product of nature itself. In this slippage, whereby the 'helping hand' of technology is both conflated with, and yet also displaces, nature, a key shift in the cultural meaning and organization of reproduction must be seen to lie. The importance of this shift is in its legitimation and naturalization (indeed legitimation *through* naturalization) of assistance to the reproductive process. (Franklin 1995: 333–34, emphasis in original)

Nevertheless, there are limits to how much meddling with nature (*tukle med naturen*) is acceptable and there is still some resistance to lending nature a helping hand. The case of egg donation and its continued prohibition is but one example. The use of a surrogate is another. The use of a surrogate (and donor eggs) by a male homosexual married couple represents a particular challenge for the Norwegian authorities, as does the use of surrogate mothers in India, whether the commissioning party is a single person (of either sex) or a heterosexual couple. Finally, those technologies that allow for the selection of embryos or fetuses are especially worrisome, for some.

In this connection, there is one more point to be made regarding the global–local dialectic of reproductive technologies. Globalization affects body politics.

That is one significant aspect of the localizing process of globally available reproductive technologies. Globalization is not just a question of importing and appropriating technologies and their accompanying knowledges; it is also, as noted, a question of the existence of those technologies elsewhere. To be more specific: Norwegian legislators are well aware of the possibilities that exist in the global market regarding options for infertile couples. They are not regulating in a vacuum. On the contrary. They also know that some of those affected by restrictions in the law will, in fact, travel abroad. And they know that they cannot stop them. For many (and not just policy makers) the global options are an argument to liberalize the law: health services should be provided equally to all, independent of economic means. One way of overcoming unequal access is by making treatments available at home. However, others find that certain forms of treatment are morally unacceptable, even unethical. That these possibilities exist elsewhere is not an argument to permit them in Norway.

From a global perspective, national boundaries are irrelevant. Nevertheless, national legislations are not. The boundaries produced by the law are at once moral and territorial. Flagging the high ethical standards that the law projects, the state becomes coextensive with this image: being a morally responsible state. It may well be that these effects are embedded in the very prescriptions of the law, that the very possibility of circumventing the law are in a sense already implied. I do not know – and I doubt that it would be easy to find anyone admitting to such an underlying morality.[1] Yet I do know that many consider the law hypocritical, of falsely claiming high principles. I also know that the expanding market for fertility services, indeed the biomedical industry, is a significant context for what is going on in Norway. It is the very fear of commercialization of these vital matters (be it eggs, embryos or wombs) that has guided the legislation because such commodification is morally untenable, the fear being that human beings are becoming a means rather than an end in themselves. Given this global (and local) context, Norway's response has been to legislate on the basis of a precautionary principle. The discrepancy between the legal norm and actual practice was one factor that spurred my ethnographic curiosity, as was the fact that the proposed regulations were subject to much public debate. In that direction, I was as much interested in what the law produces (its effects) as I was in what has produced the law (context). This double take has undergirded my approach and my analysis.

Facts and Values

I have argued that the making of biotechnology law articulates legislators' efforts to make sense of a world that is both real and imagined. Imagination plays a significant role in conceptualizing the potentialities of biotechnology. Underpinning this imagination is the very moral fabric of society. The precautionary principle translates between the language of imagination and the language of decision. This regime of caution is not just a legal strategy for managing risk but, as I have suggested, is a political strategy for protecting and projecting socio-cultural values deemed central to Norwegian society. These values are variously articulated but revolve fundamentally around kinship and the relationship between individual and society. They concern notions of personhood, parenthoods and family formation as well as equality and choice.

This legislation, I have argued, can be viewed as a site where meanings of kinship are negotiated. With regard to parenthood, it is the ascription of proper filiation that seems to be of major concern, and in this effort the fact of biogenetic relatedness is paramount, underscoring the importance of the biological principle. Not only does it serve the best interests of the child, by ensuring that a child 'knows' its origins and hence its identity, it also establishes certainty by eliminating any doubt about who the mother or father really is. Both the prohibition of egg donation and the rescinding of the anonymity clause can be understood in light of such considerations. By insisting that the mother of a child is the one who gives birth to it and that eggs removed from a woman must be returned to the same woman, Norwegian legislators uphold an idea of a unitary motherhood. By changing paternity laws and prescribing the use of known donor sperm, the same legislators contribute towards establishing a certain paternity. In both cases, it appears, the point is to avoid what is now obvious: that maternity and paternity are both uncertain categories. One effect of these efforts, I have suggested, is that in Norway, paternity is becoming more like maternity, rather than the other way round. To what extent this tendency is more generalized is an empirical question that needs to be investigated. But, the parental leave regulations granting fathers ten weeks leave, which are non transferable to the mother, is perhaps another indication.

The tendency in the law is to privilege the separate relationships – father–child, mother–child – at the expense of the relationship between parents and their children. Thus establishing proper maternity and paternity has been central and biogenetics are given a definitive quality. This is in contrast to what the involuntary childless themselves stress (and also in contrast to the prevailing attitude in the 1950s). In their reasoning about what makes an 'own child', they imagine the different forms of relatedness that are implicated in the various available procedures. One salient finding was the importance attached to the idea that each parent has the same relationship to the child (however that is perceived). This, I understood, was in order to avoid a skewing of the conjugal relation, so that the child 'not belong' more to one than the other. Thus, underpinning their notions of an own child was an idea of equality articulated as an imagined sameness in the constitution of relationships. In order to achieve this balance many of those I talked to were willing to pursue alternatives that the law did not permit. Many would also consider adoption a good option, as both parents would be equally (un)related to the child. Thus in establishing the parental relation, a child is not necessarily or only defined by its genes but through its various possible relationship to its parents. Moreover, multiple parenthood is not a preferred alternative, neither in law nor in practice. Parenthood is an exclusive rather than an inclusive relationship.

Although notions of equality and choice also inform the debates about assisted conception, these notions and the values they articulate are more explicitly prominent in the rhetoric of the sorting society in conjunction with the use and application of preimplantation diagnosis (PGD) and prenatal diagnosis (PND). Here we are confronted with yet another problem of conception. These technologies have not only provoked an informed position on the distinction between human life and human being – that is, the moral status of the embryo – they also enable selection in so far as they reveal information about the condition of the embryo or fetus. The issue, if you recall, is whether Norwegian women (or couples) should be able to choose what kind of child they want – and not whether to have a child at all. Such choice would be possible through a more liberal practice of PGD, or if PND is applied before the twelve-week limit of self-determined abortion. Thus knowledge that the application of PGD and PND imparts may lead to a sorting of individuals according to specific characteristics.

Norwegian legislators are agreed that they do not wish to be responsible for creating a society that potentially allows for systematic selection. However, they are not agreed on whether, for example, permitting early ultrasound will in fact imply creating what has been deemed in Norway a 'sorting society'. For some, the very possibility of being able to sort away undesirable fetuses or defective embryos is tantamount to accepting a society where 'there is not room for all' – in effect, a society that discriminates against disabled people. That would, in their opinion, fly against the basic principles of equal human worth, human dignity and the right to be different. Hence, the legislation that was passed regulates when and to whom certain genetic information or information about the condition of the fetus should be made available. At stake is not only the individual body but also the social body. Here, the 'part' (genetic information) stands for another whole (society). The state, in order to avoid a greater risk (that of a sorting society) forfeits individual autonomy and choice in these matters to achieve a higher social good: a humane society based on the values of the welfare state, equality for all, tolerance for difference and respect for human dignity. However, this is in law; in practice, women and couples may – and do – act otherwise.

Controversies: Contested Sites

One question that has run through these pages is: Who should have the right to decide on such vital matters as procreative practice? As is evident, in Norway, the state, through its legislation, does not consider all reproductive choice a necessary given. Neither does it consider that research should develop unhindered; nor that certain information be automatically accessible. Rather, legislation has been put in place in order to avoid certain practices, be it in respect for the order of nature or the good of society. The legislative process, I have argued, represents an important site in order to explore the types of negotiations that are going on. Moreover, in the development of this specific politics of life, ethics constitutes a productive interface where the relationship between law and biotechnology is played out and where disagreements become evident. Conception and procreation are key areas where beliefs about kinship, personhood and being human are worked out. Viewed from a certain Norwegian perspective, biotechnologies are seen as challenging the very idea of society – as they do the notion of human being. Some of these challenges – and the way they are met – have been explored in this book. As I have stressed throughout, there is no consensus on these matters in Norwegian society – and this is amply reflected in the public documents, debates and paper trail connected with the legislative processes. However, the fact that there is consistent disagreement on some of these issues is also what makes the ethnography interesting and revealing, as is the fact that over time positions change. It is precisely because assisted conception and reproductive technologies constitute a contested site that such moral indeterminacy is brought to light. It is through the controversies that different publics are mobilized and engaged, exposing implicit values and making them available for scrutiny. My intention has been to grasp that opportunity and use it constructively towards an understanding of contemporary Norway, and more generally in order to further the comparative purpose within anthropology.

I said in Chapter 1 that I would delineate a procreative universe – a sociocultural space that embraces the complex reality in which biotechnology and assisted conception are embedded. I realize that my efforts are only partially successful. There is no obvious 'place' to draw that line as there are no obvious ways to analytically disassemble and assemble aspects of social life and social process. They are *ipso facto* usually messy, convoluted and nonlinear. I have chosen to focus on law and legislative processes with a special view to the working relationship between law and biotechnology on the one hand and kinship on the other. I have worked back and forth between these, as they represent interpretative contexts for each other. Ideas about kinship influence legislation, just as legislation sets an agenda for thinking through notions of kinship. The potentialities of ARTs are inscribed in the law, as they are in people's practices. These social phenomena are mutually constituted, and the challenge for the anthropologist is to establish meaningful ways of separating them in order to grasp 'how they work'. One anchor this choice has given me is that of time – there is a temporal linearity in the development of the legislation (as there is in the development of the technologies). One law or revision is succeeded by the next. Yet, as I have argued, law is not only process, it is also imagination and local knowledge; it is a cultural artefact productive (and not only reflective) of meaning. It is both figure and ground. Moreover, law works with or through other acts of legislation, which may also shed light on the piece of legislation at hand. I have only been able to introduce some of these.[2]

In order to grasp these various aspects and their meanings and the ways they are mutually implicated, establishing the relevant contexts has been paramount: for the legislation process, for the involuntary childless, for ARTs. Context is about the way we frame the issues to be explored and interpreted; it is about how we make sense of what we observe. That context is doubly constructed, by the people themselves and by the anthropologist. This act – of contextualization – is a way of ordering our material or bringing order to a particular social reality. In this study, this has implied a historical dimension and detailed examination of the legislative process, reading this as a cultural text to make sense of how different people (or subject positions) make sense of biotechnology. In other words, it has been a question of rendering the incorporation of biotechnology meaningful in socio-cultural terms. Moreover, in order to make this 'sense', kinship – and the different meanings attached to relatedness – has proved to be a productive and relevant analytic framework for exploring the questions tied to assisted conception, as has a consistent attention to morality and the disparate articulations of significant values. These frameworks have served me in my attempt to gain insight into not only the issues raised but more importantly into the way the debates were framed – how the agenda was set, so to speak – and not least the values under dispute. I was also able to indicate the changes that were occurring. This is because, as we have seen, there have been some significant changes in Norwegian society. Some of these have been conducive to – and perhaps a precondition for – the acceptance of ARTs; other changes have been provoked by the very potentiality of these same technologies.

Legal (Un)Certainties

By way of conclusion – and to again draw attention to the indeterminacy of legislative processes, as well as the moral ambiguities that these procreative practices throw up – I turn to one more twist in the working relationship between kinship, law and biotechnology and the drama of establishing maternity and paternity. The subject is a report submitted to the Norwegian government in 2009.[3]

In Chapter 4 I discussed a presentation given by Odd Jenvin, concerning the plight of homosexual male couples who have had children by way of a surrogate. If you recall, in his talk Jenvin intimates that the practice will eventually impinge on the law, as more and more people accept that having children by a surrogate is legitimate: 'the legislative processes will be marked by this development', as he says. The attempt of policy makers to limit a person or couple's choice as to how they wish to procreate by passing restrictive legislation has, as I have shown, prompted people to circumvent the law. Operating in a global market, they move beyond the limits established by the state, pursuing options available elsewhere. In doing so they bring these procreative practices home. This occurs not only literally, in the sense that the result of the technological intervention (to use a very technical term) is inscribed in local kin relations whether these are officially recognized or not. These practices also – and perhaps more significantly – produce effects that may not only confound, but also actually subvert the current tendency to privilege biological forms of belonging over social ones. Thus, the legislation which seeks to regulate access to a procreative universe works to shift the limits as to what is possible. It may also inadvertently shift the premises as to what is permissible. And it appears that the cultural confusion that the Jenvin case amply illustrates is doing just that, as has the case of the twins born to a single mother by an Indian surrogate (with the use of donor eggs) briefly mentioned in Chapter 4.

In March 2008, the Norwegian government 'appointed a Commission to propose amendments to the Children's Act concerning [the] establishment and change of paternity and maternity. The purpose is to adapt the Act to major social and technological developments that have taken place with regard to developments in family patterns, the possibility of assisted fertilization and establishment of paternity with a high degree of certainty'.[4] The commission submitted its report a year later. The report makes for interesting reading. I am, obviously, not alone in registering confusion with regard to accepted categories of motherhood and fatherhood that the globalization of fertility treatments has caused – a confusion that is amplified by the revised Marriage Act of 2008.[5] The report is also important as it will lay the foundation for changes in the Children's Act; it is not necessarily intended to change the Biotechnology Act. Nevertheless, as is evident, these two pieces of legislation are intimately linked, and changes in one may produce changes in the other. In what follows, I will not detail the various arguments of the committee. However, I find the mandate especially revealing, as it is indicative of what the government of the day sees as problematic legal areas. Moreover, there are a few points elaborated in the report that I would like to draw attention to, as these are pertinent to some of the arguments put forth in this book.

The mandate explicitly states that the best interests of the child are to be the basis for the committee's considerations (with a view to the UN Convention on the Rights of the Child, passed in 1990 and ratified by Norway in 1991). The committee is asked to consider the significance of the biological principle in Norwegian child law and whether this principle works to secure the best interests of the child. In this connection the committee is also asked to evaluate the state's responsibility to ascertain paternity, whether the rule of *pater est* should be upheld, and if so, should it be extended to include cohabiting couples. Moreover, the committee is asked to consider situations where 'the parties have some connection to foreign countries, especially where fertilization has occurred in countries that permit forms of assisted conception which are not permissible in Norway, such as sperm or egg donation with an anonymous donor, the use of a surrogate mother or surrogate mother in combination with egg donation'.[6] In formulating this mandate, the government not only acknowledges the problems and inconsistencies prevailing regarding the actual situation (while implicitly asking for clarification procedures) but also recognizes the (irregular) practices that are occurring and the need to address these. This action alone I find significant. In addition, the mandate reflects an effort to cope with the effects of a global infertility market – an attempt to bring it home, as it were.

I want to draw attention, albeit briefly, to a few points in the report. These have to do with the biological principle, the definition of motherhood and how maternity can be transferred, and the institution of step-child adoption. All three revolve around the committee's main concern: that a child should have two parents (irrespective of how they have been conceived and irrespective of the sex of the parents) and that it is in the best interests of the child that parenthood be established as early as possible. With regard to the biological principle, the committee affirms that this principle has been the model for parenthood (in the legislation) while at the same time pointing to existing practices (such as adoption and donor insemination) that operate on a different principle. Recognizing that the biological principle and the best interests of the child are founded on different values – are, in fact, two different principles – they question whether respect for the biological principle is necessarily in the best interests of the child.[7] Thus, they challenge the current privileging of biological relatedness, opening up a more nuanced understanding of parenthood and filiation, granting nurture and care a greater space. In consequence, they also suggest reformulating provisions for paternity claims, making it more difficult to change paternity.[8] However, in so far as the definition of motherhood is concerned, they see no reason to change the already existing provision: the legal mother of the child is the woman who gives birth to the child.[9] It follows from this that a surrogate mother is therefore to be considered the mother of the child, even if she has no intention of being its social mother.[10] In Norway the only way that maternity can be legally transferred is by ordinary adoption or step-child adoption.[11] As long as the child is born in Norway, Norwegian legislation will apply. In the same vein, egg donation does not pose a problem. Although the child is conceived abroad, if born in Norway, the woman who gives birth will be regarded as its mother. The problem (as we have seen) is when the child is born abroad with the intention of it being brought to Norway. This is the case when a surrogate mother has been employed (whether or not with the use of donor eggs).

This report is (as far as I know) one of the first (in Norway) to seriously consider the question of surrogacy and its legal implications for the parties involved, including male homosexual couples.[12] One of the main questions is how to resolve the issue of maternity. How can maternity be legally transferred (for example, in cases where male same-sex parenthood is accepted in the country issuing the birth certificate)? Another way of phrasing this question is: How is maternity to be transferred to the non-biological father? Paternity can also be an issue (even if the sperm used comes from the intending father) if the surrogate mother is married and the rule of *pater est* applies (in the country where the child is born). The committee is especially concerned that any transfer of parenthood is done so that all interested parties are considered, all the more so when the child is to cross international borders and hence legal jurisdictions.

The committee pays particular attention to the fact that in many cases the intending parents conceal the identity of the birth mother. This is in their opinion not good practice. Rather, they suggest, Norwegian authorities should be obliged to ascertain and register the birth mother, to comply with a basic right of the child.[13] The authorities should also confirm that adoption is the only way that motherhood can be transferred, and that adoption implies obtaining the acceptance of the birth mother. Thus, the committee upholds the birth mother as unique (irrespective of whether donor eggs have been used), while at the same time securing the notion that the child has only two legal parents. This is first and foremost in order to create stable relations for the child, and to ensure that there is no legal opening for others – such as a surrogate mother or her husband – to claim or disclaim a relation to a child. In other words, the committee argues that Norwegian authorities should not accept documentation (a birth certificate) from abroad that grants two men parenthood to a child. Finally, the committee suggests that it is the institution of step-child adoption that should be used to transfer maternity rights. This is a more rapid procedure than ordinary adoption. Moreover and significantly, they suggest that step-child adoption be applied in cases where the child has 'been planned' by the family – that is, in cases of assisted conception undertaken abroad.[14] The legal father's spouse can apply for step-child adoption, which will establish that all involved parties have given their consent,[15] and the child will be secured 'long-lasting legal relatedness to both those persons that are intended to be their parents in the future'.[16] In this way, they introduce the notion of procreative intent as an important element in ascertaining parenthood.

With regard to paternity, it is also necessary that this be established within a legal framework that is congruent with the laws of the country where conception took place. In so far as a transfer of paternity is legally necessary (in the case of a married surrogate and where the rule of *pater est* applies) then this transfer has to be accepted by Norwegian authorities. The point is to assure that the child has the same legal father in its country of birth as in Norway. It is only when Norwegian paternity has been established that the child will have the right to Norwegian citizenship.[17]

Although the committee explicitly questions the privileging of the biological principle, and indeed suggests that knowledge of biological origin not necessarily imply changing kinship relations,[18] it is significant that they insist that the mother is the one who gives birth to the child. Thus, they too uphold an idea that *mater semper certa est* and that a child has a right to know the identity of its birth mother. However, at another level, this report feeds into a different process all together. It says something about the working relation between law, kinship and biotechnology. And it confirms that these relations are indeed complex. It is no easy matter to establish legal subjects through a series of relationships that in different ways defines this subject. Basic kinship relationships are no longer obvious; the new procreative practices confound these relations. Procreative choice – whether within or outside the confines of the law – is a reality that cannot be ignored. It is not just a question of ascribing proper kinship relations but also exclusive ones, in order to ensure that the child will have a stable environment. The report demonstrates that reproductive technologies are implicated in reconfiguring both the legal and the social landscape of kinship and that these technologies also form part of a transnational biopolitical landscape, mobilizing state authorities in Norway and elsewhere. Reproductive intent and the best interests of the child emerge as central axes around which this landscape is formed. Finally, this report is yet another indicator of the legislators' attempt to minimally keep pace 'with developments' and will most certainly spur new debates. Whether it will result in changing the Biotechnology Act remains to be seen. Whatever the outcome, it confirms that law is process, indeterminate, and always subject to revision in efforts to overcome discrepancies and make (legal) sense of the 'real' world.

Postscript

SOME NOTES ON METHODOLOGY

This book is the result of a multi-stranded – and multi-sited – research process. I have cast my net wide in order to be able to capture the complexity of the issues related to the incorporation and understandings of assisted reproductive practices in Norway. This has implied an involvement in many different social arenas and an engagement with many different sources and types of data. As mentioned in the Preface, my research was initially framed as part of a collaborative project on 'Meanings of Kinship in Norway', with a particular focus on the involuntary childless and assisted conception made possible by the introduction of assisted reproductive technologies (ARTs) in Norway. My approach was inspired by other anthropologists, who had already gone down this path, and one aim of the project was to work comparatively and provide a further contribution to the field mapped out by these earlier efforts.

However, as the research unfolded, the ethnography begged its own questions, and I was drawn in different and somewhat unexpected directions. One of these was an engagement with legislative processes, as the practice of assisted reproduction in Norway had specific and interesting legal ramifications. This engagement turned out to be particularly fruitful, and the alignment of legislative issues with kinship and biotechnology was an especially productive research area. By being drawn into the legislative process I was also drawn into considering issues that were not part of my original research scheme, such as the moral status of the embryo, the Norwegian notion of a 'sorting society', and tangentially the question of abortion and questions of surrogacy (which were not on the Norwegian biopolitical agenda when I started my research). In order to grasp the extent and meaning of these (and other) questions in the Norwegian context, a historical perspective became essential in an attempt to locate both continuities and change.

In what follows I want to draw attention to a few methodological points. I will briefly indicate the main sources of empirical material from which I have drawn my ethnography. I will also indicate (again briefly) some circumstances that I consider significant with regard to the generation of these data. To that end I address the practice of anthropological fieldwork and how certain themes

– and particular circumstances – demand that we rethink, even redirect, our methodological practices.

One salient aspect of this research is that it has been going on intermittently for many years. I started my research in late 1999, and the main part of my fieldwork and interviews was completed in 2002. However, and as indicated in the Preface, the opportunity to pursue this research beyond its initial focus was provided by two major collaborative research projects. Moreover, as this research has involved many different sites (such as parliamentary debates, official documents, public meetings, conferences, media coverage, to name but a few) I have over the years been able to pursue other dimensions of my chosen theme; indeed, I have been able to go down unanticipated paths, increasing the extent and the depth of my data, generating new perspectives while following the shifting grounds of reproductive practices. These opportunities have in part to do with the very fact that the subjects that interest me are continually evolving, and as a consequence practices and positions are changing. However, the fact that I am working in my own society – I am, as it were, continually in the field – also contributes to a kind of constant anthropological presence.

In so far as I have not shifted my anthropological gaze to other matters or other places, it has not been easy to disengage and draw the research to a final close. This I believe was fortunate as it made very evident the processual nature of the ethnography. Process appears to be a characteristic of the matter at hand, underpinning both the development of the technologies, the practices and the legislation. It has also influenced my understanding and interpretation of my data. In some ways I would say that my research is captured by Moore's term 'current history' – and an example (I think) of how an anthropologist can address 'the historical process that is unfolding right in front of her' (Moore 1987: 730).

My initial research focussed on the involuntary childless. My introduction to this arena was through the Association of the Involuntary Childless (Forening for ufrivillg barnløse, FUB; now known as the Ønskebarn). I participated in FUB's activities, primarily in Oslo but also elsewhere. These could be local public meetings organized on specific themes of interest to the involuntary childless, or day-long seminars, gathering people from different parts of the country. On some occasions I was also asked to speak. I also participated in two discussion groups organized by the FUB that met about once a month in the course of a year. In addition I took regular part in the meetings of the secretariat of the local Oslo branch of FUB. They met every fortnight (in the afternoons) to sort out practical matters – such as sending out their magazine – and business affairs, as well as planning upcoming activities. These occasions were congenial and provided an informal setting for discussing a wide range of topics not confined to the problems of being involuntarily childless: from their personal preoccupations and life situation, to issues concerning FUB more generally, to information about treatments and their reactions to policy. Through the various arenas of FUB, I met many women and some men who agreed to be interviewed in their homes. I carried out in-depth interviews (lasting from two to over four hours) with fourteen couples. These activities took place in the early evening or at weekends, mainly because of the fact that most people were at work during weekdays.

Although my study was not focused on fertility clinics as such, it was important to obtain a medical perspective. Interviews were carried out with the most prominent doctors and heads of clinics working in the field of assisted conception. They included some of the pioneers in the field as well as practising personnel.[1] These interviews were important not only because they provided insight into the procedures and terminology of assisted conception but also and more significantly because they reflected the various perceptions of medical experts on the practices involved and their implementation in Norway. Visits to fertility clinics also included interviewing laboratory personnel (who demonstrated equipment and procedures). Moreover, I have interviewed experts on DNA testing, bioengineers involved with sperm donation, and state bureaucrats implicated in these issues. I visited the sperm bank CRYOS in Denmark, where, in addition to a long conversation with the director and talking to other personnel present, I had the opportunity to interview some of the sperm donors. I have interviewed several bioethicists who have contributed to the early and ongoing debates on biotechnology in Norway. In order to gain knowledge of the bioethical questions as phrased by philosophers and bioethicists I have participated at seminars and conferences arranged on these themes. I have also attended Nordic fertility conferences and expert meetings. At such events I have often been asked to give a presentation. I note that the boundaries between my academic work (as an 'expert' contributing to knowledge of this field) and my fieldwork became blurred, as such conferences/seminars are also important sites for learning and gathering more information about the issues and the way they are configured.

I have interviewed and talked to the secretariat of the Norwegian Biotechnology Advisory Board. Over the years I have attended their many public meetings and hearings on different aspects of the applications of biotechnology, and I have also been asked to give presentations at these. These meetings (usually lasting a day, often with international participants) draw a large audience, including the scientific experts, politicians and lay people who are somehow engaged with the issues on the agenda – such as parents of children with Down syndrome or couples who have made use of a surrogate. Invariably, the agenda is put together so as to draw forth the contradictions and disagreements regarding the themes addressed (for example, on egg donation, prenatal diagnosis or surrogacy) and therefore represent an important arena of articulation of the way central arguments are phrased, and hence the context within which they gain meaning. Moreover, such meetings provided good opportunities to socialize and discuss things with different people in a more informal manner.

I have in the course of this fieldwork interviewed former (and at the time) present members of Parliament (from the different political parties) who have been central spokespersons on issues related to assisted conception and reproductive technologies. They include former ministers of health and members of Parliament who at one time or another have sat on the Committee of Social Affairs (Sosialkomitéen) in the periods when legislation was being passed. The interviews were carried out in 2001 and 2002, implying that I have not interviewed anyone involved in the subsequent revisions (but I have had access to their statements in Parliament and in the media). I have attended several parliamentary debates and also been present at some hearings.

In addition to this, I have collated a vast amount of documents. Indeed, the paper trail has at times seemed overwhelming. Tracing a legislative process implies not only studying the law itself, but also all the documents that feed into its making: parliamentary reports, White Papers, bills, hearing documents and the parliamentary debates themselves. This material represents a significant part of my data. Moreover, in order to locate such legislative events in their time and context, I have followed comments in the media (especially newspapers, but also radio and television programmes) from the 1950s onwards. The media coverage inevitably intensifies at points when legislation is being debated, and are revealing as to what is considered newsworthy as well as of interest regarding the way this is presented. The publications of the Biotechnology Advisory Board (*Gen-i-Alt*) along with back issues of the magazine that FUB published are other valuable sources of information, reflecting the shifting perspectives that are brought to bear on these issues over time. Taken together, this textual material has proved to be rich ethnographically, not just in terms of the (f)acts of legislation, but also – and especially – as it captures a mood, a time, and not least the way the agenda is set, how the issues are framed, and how this framing evolves over time.

As should be evident, my research on kinship, assisted conception, biotechnology and legislation in Norway has implied a shift from a classical anthropological methodology to a more open-ended, multi-sited approach. In my earlier research in Argentina and Mexico, I followed the traditional method of anthropology: long-term fieldwork based on participant observation, primarily within a local community. Such an approach was not feasible in this project. This has to do with the way I have focused my subject matter, the phrasing of the main problems I wish to address, my overall anthropological concerns, as well as the choices I have made regarding the type of data I found interesting and significant. These choices have, in turn, not only had implications for generating my ethnography, but have also had repercussions for the way I juxtapose, compare and create an overall coherence between the different sets of data, between my 'sites'. As there was no 'village' of the involuntary childless or any one 'place' where the contentious issues of biotechnology were located or acted out, and as there were few immediate sites of social interaction that I could access over time as participant and observer, I had to reconceptualize what constituted my 'field', which sites were relevant and how I could circumscribe them. In other words, it is the phenomena I wish to address – my thematic foci – that has guided my approach and also underpins the kinds of connections I could make. My focus has not been on social organization and everyday social life but rather on tracing different articulations of kinship, law and biotechnologies.

Let me illustrate this point by way of a contrast between this research and that of my previous work in Mexico. In Mexico I lived in a village, and place was a relevant category for the anthropologist and for the villagers. I was present most part of the day, and to a certain extent the rhythm of the villager's daily life also became mine. I was present over an extended period of time and in very diverse situations. I not only took part in daily activities, I was also able to corroborate meanings of particular cultural statements by tracing their articulation in various settings, with various (and the same) people at different times. In the Norwegian case, this was not possible. I did meet many different people in a variety of social settings, but these settings were socially and temporally disparate. In the case of the involuntary childless, I met them in very limited and specific settings and with respect to one aspect of their lives: their infertility. I went to the meetings they went to, I may have read some of the same literature, but I did not participate in their daily lives, neither at work or at home.

In contrast to the villagers in Mexico, the people I were involved with in Norway did not necessarily know each other, and did not have much of a chance to get to know me. I saw some of them only once; others I met several times. In so far as these men and women could be said to constitute a group – or community – they did so only through their mutual condition of being involuntarily childless, sharing a state of infertility and a desire to overcome it. In other words, their social interaction (if any) was also limited to that concern; this is what brought them together.[2] They were not brought together as kin, neighbours, affines, friends, colleagues, political allies or members of the same faith. Whereas in Mexico I had access to a series of relations in a variety of settings, and the resulting ethnography was varied, covering a wide range of topics, this is not quite the case for my work in Norway. My presence as a researcher was legitimated by my interest in certain issues. This holds true for the involuntary childless, as it does for the politicians, experts and others I have talked to. People were very much aware of this fact, and the material obtained through interviews with the involuntary childless and participation in discussion groups is to a certain degree narrow (though not shallow), evolving around the same topics again and again. Repetition is in fact a salient characteristic of this material.

Precisely because I did not have access to people's lives outside clearly defined settings, I had no way of tracing or accounting for the meanings of their infertility in other life situations. I was not there. Nor did I gain immediate access to other preoccupations that may or may not have to do with childlessness (although other issues surfaced in our conversations). This absence of being present in the daily flow of events or lack of other sites of interaction is a feature of my ethnography that I have had to contend with both with regard to corroborating statements made and with regard to generating socially thick descriptions. Some of the issues that were raised in the interviews I could and did also raise in other settings, to get a grasp of different reactions. Other issues that cropped up were subsequently raised in the media and by FUB. So I did have ways of knowing whether what I was being told was exceptional or not by gauging, for example the reactions and coverage of the media. However, I cannot know how the state of infertility impinges on other parts of their lives, on their social relations more generally, except from what I have been told. Nevertheless, I did have other significant sources to draw on regarding the public meanings and understandings of kinship, assisted conception and reproductive technologies in Norway.

Another difference between my work in Mexico and in Norway has to do with my presence as an anthropologist. This takes on different meanings in the two situations. Much can be said on this point, but I limit myself to the obvious – and some of its implications. In the Norwegian case I was not a stranger in the same sense that I was in Mexico. Not only do I share a certain cultural background with the subjects involved, my daily life resembles that of other women. I am recognizable, even as a researcher. But recognition goes both ways. My knowledge of my own society is also a factor that is important in establishing a common ground, as is the way I live my life. There are differences, for sure, but in contrast to Mexico, where people were curious – and sceptical – as to who I was and what I was doing and why, my interlocutors in Norway were not much interested in me. My role as a researcher (if not an anthropologist) was recognized and accepted as legitimate – and trust, so essential to our work, I believe was established on that ground. This goes for the politicians, medical doctors and other experts, as well as the involuntary childless.

One of the anonymous reviewers of an earlier version of this book questioned 'how "truthful" people are in the interviews'. Of course, I can never be sure. And some of the interviews were definitely more substantial and in-depth than others. Yet, my feeling throughout has been that the people to whom I talked have been exceptionally frank and forthcoming. Once the theme of the conversation had been set and agreed upon, there were few issues that could not be broached. This was especially the case with the involuntary childless. In addition to their reproductive histories, I was interested in covering a whole range of themes – from family and occupational history, economic arrangements, kinship relations, friendships, sexuality, beliefs, hobbies and, not least, capturing their imaginations of a future life trajectory. Moreover, it became apparent that the interview provided an opportunity to talk freely about their situation and its various ramifications, which I was to learn was not necessarily the case in other contexts. At the end of our conversation (because that was what it was, not a formal interview) I would always ask if I had left out something important, if there was something I had not asked that they had expected or the other way round, if there was something that I did ask which they found difficult or irrelevant. Invariably they would answer that the interview was not quite what they had thought it would be: it was less formal and more relaxed. The conversation did take unexpected turns, and we did broach matters they did not immediately see the relevance of (and sometimes I did not either), but they said that the conversation had been interesting and meaningful, and importantly they appreciated my coming.[3]

As to the question of truthfulness – I would argue that even if they were concealing information (and of course they were, if not deliberately; is not that nearly always the case?) this in itself does not deter from the value of what is being said. How they phrase an issue, what they wish to talk about, what they choose not to mention – these are all telling when it comes to interpreting a conversation or rendering an event. The problem or challenge for the anthropologist is to have enough knowledge in order to discern particular gaps or silences and hence query reasons why certain subjects are not broached. The fact that I too am a participant in the 'same culture', that I too am knowledgeable about Norwegian society and modes of social interaction, was in fact an advantage with regard to grasping some of the more sensitive points or what was passed over in silence.

Let me then as a final reflection return to the practice of anthropological fieldwork. My previous experiences 'in the field' are on several counts different from the way I have carried out this research. Had I not known what traditional fieldwork implies, I would not have been aware of what I was missing: the absence of more detailed and varied situations of social interaction and hence the lack of a particular kind of data that anthropologists cherish. That is one important lesson learned. That the field itself can be reconfigured and thus provide alternative routes to accessing a social reality is another. And, not least, that thick descriptions can be generated on the basis of different sources of data. Thus, the challenge was to locate other relevant sites that would yield significant data as well as establishing relevant interpretive contexts in order to elicit meanings with regard to, for example, the nature of the relationships between infertility, assisted conception and kinship, on the one hand, or the relationship between these relationships and the law on the other.

Turning to a systematic study of public policies and debates on questions related to assisted conception and the application of reproductive technologies served me well in this regard. Exploring in depth this type of material yielded a broader knowledge of the conditions within which the involuntary childless were 'framed' so to speak, while at the same time it made me attuned to the fact that these conditions did not necessarily represent the ones within which they themselves framed their preoccupations. In addition to recording their reproductive and family histories, I engaged the involuntary childless in a conversation about the different options available – even the ones not permitted in Norway. By drawing the Norwegian legislation into the framework and asking questions about limits, ethics and what they themselves would (potentially) be willing to do effectively contributed to my understanding of their notions of kinship and meaningful relatedness. In other words, the legislation served to contextualize practices I encountered among the involuntary childless. Moreover, and significantly, these public policies and legislative events became ethnography in their own right – they directed the thrust of my subsequent research. Thus the legislative process serves at one and the same time as both figure and ground.

In my attempts to find ways of establishing relevant contexts through which I could inscribe and interpret some of my data, I simultaneously generated more data. I compounded the ethnography – and created some disjunctions. It may well be that these disjunctions are not just – or only – a product of my research, but do in some ways adequately capture dimensions of the social reality I have attempted to depict. It is not only a question of law and practice. I also have in mind the public/private interface and the way this plays out in the Norwegian setting, as well as questions of filiation and biogenetic relatedness, and the relation between individual and society. On the one hand, the fact that that the state intervenes in intimate spheres of social life has implications for the way that life is lived and perceived. So does the way that the media makes intimate medical conditions a matter of public, at times even sensational, interest. Hence, to juxtapose the lives and experiences of the involuntary childless with the actions of policy makers (or the press) is to render relevant the social order within which these experiences gain their meaning. On the other hand, in order to grasp the overarching significance of the way the involuntary childless perceive their infertility and the possible application of assisted reproductive technologies, and what this in turn means for the perceptions of these technologies more generally, I had to inscribe their meanings in a wider context. I too have framed my subjects, and it is my way of contextualizing – of rendering the ethnography meaningful – that underpins my anthropological perspective. In fact, this research has forced me to address the problem of context and acts of contextualization in ways I have not done previously.

Nevertheless, the fact remains that disjunction is what characterizes this field situation and the data I have gathered both with respect to time (a lack of continual presence in one setting yet carrying on over a number of years) and with respect to the kind of data collected. To a much greater extent than in my previous research, I have turned to texts as an important source of data. This has, as mentioned, included collating and systematizing a wide array of textual sources in order to capture a sense of the public and to be able to align and relate these to the private, intimate and personal experiences of the involuntary childless. Not only do these texts constitute data of a different kind, they are also disparate, of different orders and kinds. A legislative act is not of the same order as a media event reporting on a surrogacy arrangement in India. Yet, as I hope I have showed, they are linked, sometimes only implicitly and sometimes explicitly. One major challenge has been to draw these different empirical sources, the public and privates spaces, together in order to establish some coherence, some circuit of meaning. To establish this coherence implies at one and the same time a decontextualization and a recontextualization. At a more general level, I would suggest that what brings all these sites – or social spheres – together are different social imaginations. These imaginations – emanating from different subject positions – work together to produce ideas of society, family, kinship and the individual.

Working in a society that I am a part of and within which I am also inscribed is an advantage on many counts, but it does also require another kind of sensitivity and alertness so as to avoid the pitfalls of ethnocentrism. To a certain extent I shared some of the socio-cultural premises of my interlocutors. I shared some overarching values. Thus, to avoid assuming any common ground, I had to be especially reflective and create the necessary critical distance, always questioning how and in what ways I could best explore and bring forth interesting phenomena and significant observations. I believe my previous fieldwork has also served me well for those purposes. In addition, the fact that I am writing about my own society in a foreign language contributes significantly to creating this critical distance. I cannot assume any common ground with my readers, and therefore am obliged to render my ethnography in terms that not only make sense to an 'outsider' but also appear as plausible. This is not just an act of cultural translation – it is in fact an important methodological point. One of the operating principles that underpins anthropological methods is a practice of distanciation: making the strange familiar and the familiar strange. This form of distanciation is deliberate – a heuristic tool – and is perhaps more explicitly present the closer you are to 'home'. In my case, there is no doubt that depicting in English matters concerning my own society has effectively contributed to a critical reflexivity.

The craft of anthropology is based on methods that give us access to the natives' point of view – that is, an understanding of people in their own terms. In this case, the indigenous view is brought forth through the involuntary childless and through an examination and analysis of the Norwegian *offentlighet* (public domain), especially as reflected through a legislative process, but not only. By framing my work in terms of kinship and relatedness and consistently focusing on the way reproductive technologies are inscribed in different practices – which includes the making of a law as well as different procreative acts – has led me to examine a variety of sources, accessing different sites. Anthropology always works at an interface, and consists among other things of bridging different or even inter-penetrating life-worlds – of translating between one and the other, of demonstrating connections – whether in our own societies or elsewhere. This is the work we do. In this work the anthropologist represents a kind of intervening space – an interstice. The way I have set my themes and delineated this procreative universe represents a contribution to understanding significant dimensions of how this social reality is constituted. It will also hopefully serve its purpose in future comparative work, be it to social realities in Norway or elsewhere.

Appendix

FERTILITY RATES, TRENDS AND POLICIES IN NORWAY

The significance of being a family – understood as having children – should also be seen against the background of fertility rates in Norway. The National Bureau of Statistics reports that 2008 saw the highest fertility rate in Norway since 1975; it had then reached 1.96. The average birth age for mothers for a first born child was 28.1 over the previous four years; the average birth age for all births was 30.3. The average birth age for fathers was 33.4 and had been stable for the past six years.[1] Lappegård (2007: 56) notes three major changes in fertility patterns in Norway from 1960 until the present:

1. The total fertility rate has decreased from about 3 to 1.8 per woman from the middle of the 1960s to the present.
2. Over the last few decades, a general postponement of parenthood has occurred. In the 1960s and early 1970s women would on average have their first child at 23; today that average is 28. However, that women postpone childbearing does not mean that they are not having children. Women of 30 to 34 years of age today give birth to 60 per cent more children than women in the same age group did in the early 1970s. This pattern of postponement and 'catching up' (*gjeninnhente*) characterizes the Nordic countries, in comparison with other European countries.
3. Over the last few decades there has occurred what is termed a 'destandardization' of the fertility pattern. This implies that there is greater variation in the number of children that each woman has.

At the same time, changes in marriage and cohabitation patterns have occurred. Women marry later (30 years today as against 23 in the 1970s). Cohabitation has become a normal part of the lifecycle. More people are living alone, which is explained by the increased rate of divorce and separations, while at the same time more people have several relationships over time (what is often termed 'serial monogamy'). Twenty per cent of 35 year olds today have experience of at least two stable relationships, compared to 3 per cent in the same age category twenty years ago (Lappegård 2007: 57).

Lappegård explains these changes along three dimensions: the contraceptive revolution, giving women a real choice with regards to fertility (whether to have children or not, and how many); shifts in women's individual autonomy; and women's liberation. Women's autonomy has increased with increased education and increased participation in the labour market.

Thus, while fertility has fallen to low levels in many industrialized countries since the 1980s – for example, Italy and Spain have a fertility rate of 1.3 – rates in Norway have been on the rise. At the beginning of the 1990s the total fertility rate had been just below replacement. At the same time, postponement of motherhood to a later age had been steadily increasing. The proportion of women giving birth for the first time between the ages of 30 and 35 had increased from 5 per cent among the 1950 cohort to 10 per cent among those born ten years later (Lappegård 2000).

Lappegård (2000) explains the tendency to postpone first childbirth through increased educational levels. The proportion of women completing college or university education has doubled since 1980, and in 1996 more than 20 per cent of all women over the age of 16 had higher education. The proportion of childless women was highest among those with the highest education level. Nevertheless, Lappegård suggests that changes in family policy in the 1990s have made it easier to combine participation in the labour market with having children.

Norwegian gender equality policy is based on women's education and participation in the labour market. Women's participation in the labour market has not implied that that they do not want children. As Lappegård states:

> The family institution is strongly embedded in the Norwegian society. Surveys indicate that women want a family, but that they also want to participate in the labour force ... Younger female generations have grown up in a time when gender equality and new patterns of family formation have been established. From a life course perspective these women probably consider labour force participation as natural as child-raising. (Lappegård 2000)

From this perspective it is also notable that Norway has a high level of occupational sex segregation – in fact, it is claimed that the Norwegian labour market is one of the most sex segregated in Europe. This has been coined as the Norwegian gender equality paradox: a combination of a high degree of gender equality in society with a gender-segregated labour market.

According to a government report, this paradox should rather be viewed as a dilemma for the Norwegian welfare state.[2] The authors claim that there is no doubt that the high level of employment among women in Norway is due to the provisions provided by the welfare state, and that unfavourable comparison with other societies might overlook the unpaid domestic labour that women carry out in countries which have no such provisions. Nevertheless, researchers have pointed out that the welfare provisions provided in Norway are gender conservative (*kjønnskonservative*), and rather than furthering gender equality they contribute towards a maintenance of a gender-segregated labour market. From 1970 until today, women's participation in the labour market has increased from 45 per cent to 70 per cent. However, 44 per cent of women work part time, as compared to 13 per cent of men. Of these, many are employed in the public sector. From 2002 onwards, Norway has had the highest number of employees in the public sector within the OECD area. In 2005, 47 per cent of all employed women worked in the public sector, whereas only 18 per cent of men did the same. Since the middle of the 1980s women have received more higher education than men. However, there is a marked difference in the choice of educational training. Women predominate within teaching, health and social services whereas men opt for natural sciences and technical training. The report concludes by suggesting that the gender-segregated labour market can in part be explained by some of the state's welfare provisions – such as parental leave and cash support (provided for those, mainly women, who stay at home and take care of their children) – which tend to strengthen women's role as mothers.

Lappegård (2007: 60–65) also discusses the relationship between work and family life. She mentions two mechanisms that seem central to the positive relation between female participation in the labour market and fertility: access to child care and welfare state family policies. One central point is whether these policies increase or decrease a woman's dependence on the family.

Norwegian welfare-state policies seek to reduce the tension between the demands of work and family, and work as a buffer between the labour market and parenthood (Lappegård 2007: 61).[3] Some examples of these policies are: generous maternity and paternity leave,[4] access to state-supported kindergartens, a regulated labour market, the right to unpaid care leave (*omsorgspermisjon*) for parents of small children, and the right to paid leave to take care of sick children. To increase gender equality it is deemed important that men be drawn into the family sphere. The Norwegian state has, therefore, as part of its parental-leave policies instituted a 'father quota'. In 1993, Norway was the first country in the world to grant fathers four weeks of paternity leave. By 2009 this had been extended to ten weeks. These weeks cannot be transferred to the mother and are lost if the father chooses not to take them. There is also cash support available to parents of small children.[5] Cash support is given for all children between 1 and 3 years of age who do not or only partially make use of kindergartens or day-care centres that receive public support. This cash support is given to the parent with whom the child lives permanently. The cash support comes in addition to child welfare (*barnetrygd*) which is given for all children under 18.

Notes

Preface and Acknowledgements

1. Unfortunately, Olaf Smedal was not able to realize his empirical research but was, nevertheless, an active partner in the initial project. In addition to engaged discussions and the supervision of master students enrolled in the project, he has left his mark through his thoughtful introduction to our edited volume, Howell and Melhuus (2001).
2. The Norwegian Research Council, through the anthropology committee, first contributed with some initial seed money in 1997 to initiate and formulate a project. In 1999, we received a larger grant under the rubric 'tiltak for å styrke forskningsprosjektenes andel av fri prosjketstøtte', which was an effort to increase financial support for 'free research' – that is, research that is not part of a larger pre-designed research programme.

Chapter 1

1. Biotechnology is the term used in Norway to denote those procedures and technologies that regulate assisted conception and associated practices. This is reflected in the very title of legal documents. In shorthand it is referred to as *bioteknologiloven* or the Biotechnology Act. I follow local usage of this term, though I am aware that biotechnologies also have a much broader reference.
2. The Postscript provides reflections on methodology, on fieldwork 'at home' and the generation of relevant ethnography, including considerations about the use of documents and texts. For those who wish to gain insight into the background of this research, I suggest reading the Postscript first.
3. In Norway, biotechnology was initially separated into two areas of application: one which concerned gene technology in relation to plants, animals and the exterior environment, and one which concerned gene technology in relation to humans (see Brekke 1995: 9–11). Thus, already from the start the human/non-human divide was iterated. This has had repercussions for ethical concerns and debates. It has also resulted in two different legislative acts regulating biotechnology: the Gene Technology Act (Lov 1993–04–02, nr.38) and the Act Relating to the Application of Biotechnology in Medicine (Lov 1994–08–05, nr.56), passed in 1993 and 1994 respectively. Full Norwegian names of these and other acts are given in the bibliography.
4. The Artificial Procreation Act (Lov 1987–06–12, nr.68).
5. In legal terms, the Norwegian Biotechnology Act (Lov 1994–08–05, nr.56) is classified as *lex specialis* in contrast to *lex generalis* (Rommetveit 2005). This implies that in cases of conflict with a more general legal framework (such as that of patients' rights) the special law is given priority. Rommetveit observes that the Biotechnology Act can be characterized as based on prohibitions rather than being based on rights (Rommetveit 2005: 169).

6. See Delaisi de Parceval (2008: 377, annex), which provides a table covering practices of ART in twenty countries.
7. The term 'artificial' is no longer used with regards to insemination, though this was the common term. I do not specifically address ICSI (intracytoplasmic sperm injection), a micromanipulation procedure where a single sperm is injected into an egg to achieve fertilization, nor GIFT (gamete intrafallopian transfer) and other of the possible technologies used in combination with IVF. However, ICSI has become a common procedure in Norway. See Inhorn (2007) for a discussion of the significance of ICSI in Muslim countries.
8. Act Relating to the Application of Biotechnology in Medicine (Lov 2003-12-05, nr.100).
9. For a brief overview of fertility rates, trends and policies in Norway, see Appendix.
10. In her succinct introduction to the edited volume *European Kinship in the Age of Biotechnology*, Edwards (2009) takes issue with the so-called new kinship studies, problematizing the very idea of 'new'. She questions whether 'new' means a change in epistemology or a change in practice, as well as the relationship between the biological and the social, asking whether (and when) 'the biological' is understood as (necessarily?) the same as nature.
11. Of course it has not only been anthropologists who have worked on issues provoked by the new reproductive technologies, though the focus on kinship relations can perhaps be said to be a more specific anthropological concern. Some of the early texts are Stolcke (1986), Stanworth (1987), Haimes (1988, 1990), Spallone (1989), Strathern (1992b), Fox (1997) and Edwards et al. (1999). See also Holy (1996) and Bestard (1998).
12. The exception here is Schneider's seminal study of American kinship, first published in 1968 (Schneider 1980), and his subsequent critique of kinship studies (Schneider 1984). These studies have served as a benchmark for the reconfiguration of kinship studies, including perspectives on gender and power. See, e.g., Strathern (1981, 1992a), Collier and Yanagisako (1987), Yanagisako and Delaney (1995) and Franklin and McKinnon (2001b). Moreover, both Strathern (1999) and Konrad (2005) draw on Melanesian ethnographies to draw forth relevant comparisions.
13. See, e.g., Strathern (1992b), Ragoné (1994), Ginsburg and Rapp (1995b), Franklin (1997), Franklin and Ragoné (1998b), Edwards et al. (1999), Carsten (2000), Finkler (2001), Franklin and McKinnon (2001b), Inhorn and van Balen (2002), Thompson (2005) and Porqueres i Gené (2009). For a recent overview, see Inhorn and Birenbuam-Carmeli (2008).
14. Rabinow's book (Rabinow 1999) is interesting in this regard precisely because it brings together a North American and French perspective on notions of the body. For more detail on the French situation (with regard to reproductive technologies), see Delaisi de Parseval (2008). Meanwhile, for Italy, see Bonaccorso (2009); and for Scandinavia, see Lundin (1997), Tjørnhøj-Thomsen (1998) and Adrian (2006).
15. For detailed analyses of the meanings of embryos and fetuses from a cross-cultural perspective, see Morgan (1992, 1998); see also Morgan (2002) and Kaufman and Morgan (2005). Questions relating to the status of the embryo are also reflected in the growing field of bioethics. For a comparison of Germany and Israel, see Hashiloni-Dovel (2007). See also Kerridge et al. (2010).
16. Geertz's essay was written against another background in anthropology, at a time when other battles in anthropology were at the fore. The thrust of his argument concerns how to do a comparative study of law; this, he says, should consist in cultural translation rather than being an exercise in institutional taxonomy or a celebration of tribal instruments of social control (Geertz 1983: 219). See also and Starr and Collier (1989), Merry (1992) and Fuller (1994).
17. A focus on litigation and dispute settlement was central to the early development of legal anthropology, especially (but not only) in local, micro-level studies of social organization. For a review of the literature and trends since 1975, see Merry (1992); see also Dolgin (1997).
18. The involuntary childless are defined as those couples who have had regular intercourse (twice a week) without using contraceptives throughout a year without conception (Sundby 1989; n.d.). See also Sundby and Guttormsen (1989), which was one of the first books to treat the question of involuntary childlessness in Norway.

19. The first Norwegian adoption law was called 'To Forge a Link' (Lov 1917). It was nine years ahead of the first English legislation and amongst the first in Europe, and came in the wake of more than two decades of codifications of various laws that resulted in the Castberg Children Acts (Melhuus and Howell 2009). A central concern was the plight of illegitimate children, in Norwegian termed *uekte*, which could literally be translated as 'not authentic', alluding to being born out of wedlock (*ekteskap*, the Norwegian term for marriage).
20. For the original Act and its revisions, see: Lov (1994–08–05, nr.56), Lov (2003–12–05, nr.100), Lov (2007–06–15, nr.31).
21. Pottage's usage of the term biotechnology is much broader than mine. He is not primarily concerned with reproductive technologies. However, his arguments are equally relevant for my more limited scope. See Pottage (2007).
22. For a different twist on the futures argument, see also Franklin and Lock (2003).
23. In February 2011 the Directorate of Health submitted its report (commissioned by Parliament) on the evaluation of the Biotechnology Act: see Helsedirektoratet (2011).
24. 'Kong Haakon var ikke Olavs far?, *Dagbladet*. Retrieved 14 October 2004 from: http://www.dagbladet.no/?/nyheter/2004/10/14/411285.html.
25. 'Was a King of Norway Really Made in England?' *Daily Mail*, facsimile reproduced in *Aftenposten*, 16 October 2004, p. 2.
26. Laura Peek, 'The British Royals of Norway', *The Times*, 15 October 2004.
27. See Melhuus and Howell (2009) for a more detailed examination of this event.
28. Strathern continues: 'Information about kinship can be used to regulative ends … But the material also shows the constitutive nature of procreative facts in the recognition of relationships. This kind of knowledge has particular resonances for Euro-Americans. Another way of putting this would be to say that "biological" information has immediate (simultaneous) "social" effect' (Strathern 1999: 75).
29. See Cadoret (2009) for an elaboration of this argument.
30. The term Nordic is generally used to refer to the Nordic countries: Denmark, Sweden, Norway, Iceland, Finland. Some commentators also include the Faeroe Islands, Greenland and Åland islands. Scandinavia, meanwhile, is used to refer to Sweden, Denmark and Norway (see Sørensen and Stråth 1997a: 19–24). However, authors are not necessarily consistent – nor precise – in their use of these terms. I use Nordic in line with Dahl, and Scandinavia to refer to Denmark, Norway and Sweden.
31. My translation is not literal. The original Norwegian reads: *Det er i denne sammenheng nyttig å skille mellom* likhet som premiss for sosial samhandling *på den ene siden og* likhet som regulerende prinsipp for velferdspolitiske institusjoner *på den andre* (Vike, Lidén and Lien 2001: 16, emphasis in original).
32. See, e.g., Gullestad (1985, 1989: 109–22, 2001, 2002, 2004).
33. See Barnes (1954) for an early observation of such processes.
34. Kildal and Kuhnle state that 'the hallmark of the contemporary Nordic "institutional" welfare state is aptly expressed in terms of three essential features: social policy is *comprehensive*; the social entitlement principle has been *institutionalised*; and social legislation has a solidaristic [sic] and *universalist* nature' (Kildal and Kuhnle 2005a: 6, emphasis in original).
35. It is worth noting that the editors conclude their introduction by stating that 'the Nordic Enlightenment tradition is in crisis for the first time', and that this 'crisis is not just economic but also cultural' with an outcome which is highly uncertain (Sørensen and Stråth 1997a: 24). Some of these aspects are the theme of another edited collection: *Normative Foundations of the Welfare State: The Nordic experience* (Kildal and Kuhnle 2005b). These edited collections all have the Nordic model as their focus, and although they do differentiate between the Nordic countries they also draw out the similarities. For my purposes, I draw on their generalizations to indicate factors relevant for Norway.
36. But see Kildal and Kuhnle (2005a: 13–15) for a more nuanced discussion of 'universalism'. See also Kuhnle (2001).

138   Notes

37. Such legislation may include the right to self-determined abortion, the right to maternity and paternity leave, and the right to health care, to name but three.
38. The Norwegian Constitution is available at: http://www.stortinget.no/en/In-English/About-the-Storting/The-Constitution/The-Constitution/. Retrieved 15 March 2011.
39. The fact of a state Church is not in itself a criterion for classifying a state as secular or not. A state Church model differs from confessional states in that it in principle secures freedom of religion and confession for all individuals and groups, and by the fact that as a matter of principle there is a divide between the state and the Church's areas of responsibility and competence. This ideal type is in most cases 'impure', and such is the case for the Norwegian one (see Plesner 2005; Bangstad 2009). It is beyond the scope of this book to trace the relationship between Church and state, which goes back to the Reformation and the specific Lutheran regimes regarding divine and state authority. For a condensed version of this history, see Brekke (2002).
40. Gullestad also argues that many Norwegians implicitly mobilize Christianity as a central dimension in 'our' imagined moral community (Gullestad 2001: 56).
41. This is a point I derive from my own ethnography, but is also made by Stråth. He states: 'Samhälle, Gesellschaft, society, societas even became synonymous with the state in the twentieth century. When the social democrats argued for the welfare state they talked about the need for a strong society. Samhälle was never in opposition to the state or seen as something between [the] market ... on the one side and the state on the other' (Stråth 2005: 36).
42. The significance of autonomy as a more embracing value is also evident in the fact that Norway has twice voted, in national referendums, against joining the European Union (1972 and 1992).
43. For an interesting example related to food practices, see Døving (2009).
44. Strathern continues: 'The several commissions of inquiry into new reproductive and genetic technologies set up since the early 1980s indicate the extent to which governments consider this a field where they should marshal public opinion, while in civil disputes litigants use whatever they can in support of their arguments. Their cultural explorations often push to the limit assumptions about what constitute parent–child relations, in the very act of seeking limits or establishing reasons for claims' (Strathern 1999: 65).

Chapter 2

1. Since I started my fieldwork, FUB (see http://www.fub.no/) has changed its name to 'Ønsekbarn: Forening for fertilitet og barnløshet', or 'Desired Children: The Association for Fertility and Childlessness' (see http://www.onskebarn.no/). In addition to its statutes and membership rules, the association's home page includes a description of the organization, its vision, mission and work. The association has several local groups throughout Norway. They organize public meetings, group discussion sessions, give advice and also lobby vis-à-vis Parliament. To ensure anonymity, all names used in this chapter have been changed and some biographical details have been altered.
2. The anthropologist was Sidsel Roaldkvam.
3. When I started my fieldwork, ICSI was not an accepted practice (see Chapter 1, note 7). Since then, this intervention has also become routine.
4. This attitude is also confirmed by Howell (2003) in her work on transnationally adopted children in Norway.
5. Obviously, many of those who choose to adopt may have no other choice if they want to have a child. Nevertheless, I suggest that the arguments used to legitimate choosing adoption are relevant whether it is one among other alternatives or the only alternative.
6. I received letters and cards from several of the women that were in treatment when I began my research, who at a later date got pregnant and gave birth to a child. They invariably included a photograph of the newborn baby.

7. In theory, the adoption agencies demand that couples have terminated infertility treatment before they put in their application for adoption. In practice, couples have these various options on the table quite early on in the process, some even riding two horses at the same time.
8. Since I began my research this situation has also changed. Surrogacy was barely on the agenda at the time I carried out my interviews. It is currently a major political issue. The question of surrogacy will be pursued in Chapter 4.
9. The talk groups were organized by FUB and were meant to bring together people who wished to talk about their various experiences of infertility. I had the chance to participate in two such groups (one involving couples and one involving only women) that met intermittently in the course of a year. In contrast to the interview situations where I would set the agenda, I tried to be more observer than participant in the talk groups. This strategy was not always successful, however, because as often as not I was asked to offer my opinion.
10. I should add: almost all those I talked to were in favour of anonymous sperm donation. This may have to do with the fact that this is what the law in Norway at the time prescribed. Yet the people I talked to were aware that Sweden had rescinded its anonymity clause and that this was also being proposed in Norway. In 2003 the law was changed, prescribing the use of sperm from known donors and granting the children born using this procedure the right to know their biological origin at the age of 18 (see Chapter 3).
11. See Howell (2006: 27–31) for a description of the adoption process in Norway.
12. This conversation took place before the law rescinding the anonymity clause had been passed. I do not know how couples today reason, nor whether there is still a preference for anonymous sperm (which would imply that the woman/couple must travel abroad to get access to anonymous sperm; no record is made of such cases). However, according to news reports, there is a scarcity of Norwegian donors; Norwegian sperm banks thus have a greater demand for sperm than they have supply, and public clinics have had to suspend treating new couples (see the report by NRK news: http://www.nrk.no/nyheter/1.7519093, retrieved 15 March 2011).
13. The public health system allows for three IVF attempts. If they are unsuccessful and the couple wishes to continue with IVF treatment they have to go to a private clinic and pay the costs themselves.

Chapter 3

1. In 1987 Parliament passed the first law regulating assisted reproduction in Norway: Artificial Procreation Act (Lov 1987–06–12, nr 68).
2. See also Franklin, who, in her discussion of the debates concerning the embryo in Britain's Human Fertilisation and Embryology Bill, argues that: 'parliamentarians were faced by a need for certainty: for guidelines, boundaries, facts, definitions, and above all *limits*. Like nature, society will not tolerate a vacuum. Parliament has succeeded in its aim, by enacting laws, to fill the "legislative vacuum" surrounding embryos' (Franklin 1999: 162–63).
3. Illicit sexual relations (whether real or imagined) are often the subject of rumours, which in turn may represent interesting ethnographic data, disclosing core values (Melhuus 1997). However, should a child ensue, the event invariably becomes public in some sense. Questions of paternity and the presumed legitimacy of a child are examples where gossip can run rampant.
4. Other procreative issues that have, in Norway and elsewhere, drawn ample public attention have included the struggle to make contraception universally available, the legalization of abortion and the granting to women of self-determination in matters of reproduction.
5. Yet, as we have seen, the involuntary childless themselves differ as to how open they feel they can be about their problems with conceiving. Their openness is one way of making their condition public. However, this 'public' is restricted to their immediate relations. Nevertheless,

individual cases are often the stuff of media coverage, which is one way that these personal dilemmas are made known to a broader public. Intimate medical details contribute to this knowledge.
6. Obviously, the laws regulating assisted conception are not alone in mirroring shifting social concerns. Other legal codifications are also productive in this regard (such as naming laws, marriage laws, paternity laws, abortion laws, and so on). However, as my focus here is on the laws that regulate assisted reproduction and associated technologies, tangential laws will only be mentioned when they bear on the issues at hand.
7. See Lov (1987–06–12, nr 68). The British government's Human Fertilisation and Embryology Bill was introduced in 1989 and enacted in November 1990. However, in 1982 the British government had already appointed the Committee of Inquiry into Human Fertilisation and Embryology – the so-called Warnock Committee – which submitted its report in 1984 (subsequently published in 1985). Thus, these debates were well underway in Britain when Norway decided to legislate.
8. These were: Byråsjef Carl Stabel, sogneprest Johannes Smidt, reservelege Hans Sundfør, dosent Åsa Gruda Skard, and housewife Siri Thoresen (Innst. 1953, p.3). With regards to the committee's composition, the government (which was then Labour) reasoned that it needed a legal expert, a medical expert in gynaecology, a representative of the state Church (due to the religious and ethical questions) and finally two women, one being a recognized psychologist.
9. The Director of Health in Norway at the time (1948) stated on the basis of information gathered from eight practising gynaecologists that at least thirty-two instances of AID had occurred in the country; that demand was increasing; and that many doctors hesitated because of legal insecurity. It is interesting to note that the Director of Health also suggested that 'AID should be free (of charge) and not only limited to married women' (Innst. 1953, p.5). This last suggestion was not put forth again until 2007, when it was included in proposed revisions to the Biotechnology Act. He also stated that in the case of married women, it is doubtful whether the husband's written consent was necessary in all cases; if the wife wants to make use of AID and the husband is opposed, her 'natural' right (to have a child) is co-extensive with that of society (Innst. 1953, p.5).
10. Similar committees had been appointed in Denmark and Sweden and they had already initiated cooperation with the aim of establishing the same set of rules. In December 1948, a meeting discussing the issue was held in Stockholm, at which Norwegian representatives from the Ministry of Justice and the Ministry of Health were also present; they were later appointed to the committee (Innst. 1953, pp.3, 5). Efforts at joint legislation between Scandinavian countries had already proved successful with regard to marriage laws (Bradley 1996; Melby et al. 2006).
11. See 'Innstilling fra innseminasjonskomitéen' (Innst. 1953). In what follows I quote and paraphrase this report to some extent; all translations are mine. Moreover, in translating I have tried to stay as close to the original as possible, even though this implies using terms and expressions that might not comply with standard English. Nevertheless, the way things are formulated does give a feel of the zeitgeist.
12. The principle of *pater est quem nuptiae demonstrant* (literally, 'marriage indicates whom the father is') implies that the father of the child is the man who is married to its mother. This principle is a basic assumption of European thinking about filiation, ensuring the legitimacy of offspring, and hence also distinguishes between legitimate and illegitimate children. This principle is also one that has been challenged, a point that will be discussed in Chapter 4.
13. Innst. (1953), p.15.
14. In Norwegian, 'nature and nurture' is phrased as *arv og miljø*, which literally means inheritance and environment, and alludes to genetically inherited traits and those traits that are influenced by the socio-cultural environment.
15. Innst. (1953), p.16. For an elaboration of the relationship between the adoption laws and laws regarding assisted conception, see Melhuus and Howell (2009). For other discussions of AID, see Daniels and Haimes (1998), Blyth and Spiers (2004) and Hargreaves (2006).

16. Innst. (1953), p.50.
17. Innst. (1953), p.17.
18. In 1945 child benefit, payable directly to the mother, was introduced to help cover the costs of children's upbringing (Leira 1992: 102). Moreover, Leira notes that 'Norway is the only Scandinavian country which has instituted a special entitlement for single providers ... This allowance has made it possible for unsupported unmarried, widowed or divorced women to have a small income of their own, which may be one reason why very few babies born to Norwegian single mothers are given up for adoption' (Leira 1992: 102). Leira is concerned with the development of child-care institutions in Norway (in comparison with Denmark and Sweden), and she demonstrates how Norway differs. In contrast to the other Scandinavian countries, institutionalized child-care was not the result of a drive for encouraging women's employment but rather based on an idea of the best interests of the child. As she says: 'In Norway, the wish to preserve the domesticated-mother family apparently was stronger than were the demands of the economy' (Leira 1992: 104).
19. For a history of Norwegian marriage laws, see Melby et al. (2006); see also Melby et al. (2000).
20. Innst. (1953), p.23.
21. Rønne-Pettersen's book was titled *Provrörsmänniskan: En studie i moderne magi* ('Test-tube Human: A Study of Modern Magic').
22. Løvset was employed at the University of Bergen. He had administered an investigation into patients' attitudes towards AID. He sent out a questionnaire to 1,200 persons, where 395 answered. Of these, he reports that only 19 were against AID. The article presents some of the reasons that the women respondents give for their position. See also Molne (1976).
23. It is therefore interesting to me as an anthropologist that the foreword to Rønne-Pettersen's book was written by Guttorm Gjessing, the then Director/Chair of the Ethnographic Museum in Oslo.
24. Innst. (1953), p.17.
25. Innst. (1953), p.54.
26. Innst. (1953), p.53.
27. Innst. (1953), p.56.
28. Innst. (1953), p.56.
29. Innst. (1953), p.55.
30. However, in 1972 the Women's Clinic at the Rikshospitalet (the National Hospital in Oslo) in consultation with the Directorate of Health initiated a programme for AID, which was started in 1973. By 1976 the hospital's capacity to carry out this treatment was reached (Molne 1976).
31. Another explanation given was the fact that the Nordic initiative to coordinate policies failed, but again it is not clear to me why this was so. It seems plausible that it was the very controversies surrounding AID that were a primary factor.
32. Kåre Molne (professor and MD) is a key figure in this field. As a medical doctor he was in charge of developing clinics which offered treatments for assisted conception, first at the National Hospital in Oslo with AID, and later at the Regional Hospital in Trondheim. He was also directly involved in the formulation of the legislation, so much so that his name was referred to in the parliamentary debates.
33. Lov (1987–06–12, nr 68). The Norwegian title of the Act is 'Lov om kunstig befruktning': 'artificial procreation' is the official translation of *kunstig befruktning*, which literally means 'artificial fertilization' or 'artificial conception'. In 1994, 'assisted procreation' was substituted for 'artificial procreation'. When referring to the legal documents I use the terms referred to in the documents.
34. Innst. O. nr.60 (1986–1987), p.2.
35. See Halvorsen (1998), whose arguments tend toward the latter.
36. The following is my translation of the Act (Lov 1987–06–12, nr 68), and I have purposely used the terminology applied in the legal text.
37. I use the term 'fertilized egg' – and not embryo – as this is the term generally used in Norway, and employed in these documents.

38. My interviews with members of Parliament's standing Social Committee active in the period from 1987 to 1994 and other members were carried out in 1999 and 2000. The politicians were retrospectively recalling an event and a period earlier in their careers, and their recollections were most likely influenced by succeeding events. Some might even be commenting on both the acts of 1987 and 1994. According to Sirnes, however, the debates in 1987 and 1994 were very similar (Sirnes 1997: 223), indicating that a possible conflation is not surprising.
39. Even in Parliament several speakers made it clear that although the proposed law may seem liberal to some Norwegians, in an international context it was very restrictive.
40. This is in contrast to the Gene Technology Act (Lov 1993–04–02, nr 38), where according to Brekke the experts were able to dominate the field, including with regard to ethical questions (Brekke 1995: 99, 107). However, over time, there has been an increasing involvement of public opinion in ethical questions in relation to, for example, genetically modified organisms (GMOs).
41. 'Research' is the term used and it subsumes the idea of science and developments in science.
42. I will not detail the positions taken by the different political parties. Suffice it to say that overall the Labour Party (*Arbeiderpartiet*), the Conservative Party (*Høyre*) and Progress Party (*Fremskrittspartiet*) are the most positive towards biotechnologies and their applications, while the Christian Democratic Party (*Kristelig folkeparti*), Centre Party (*Senterpartiet*), Liberal Party (*Venstre*) and Socialist Left Party (*Sosialistisk venstreparti*) are more sceptical, if not negative. However, within this constellation there are variations that cut across party lines. Significantly, feminists aligned with Christians in arguing for restrictive legislation. See Brekke (1995) for a more detailed account of the various positions, and Sirnes (1997) for a detailed analysis of the debate on research on embryos in Norway and the UK.
43. See St.meld. nr.25 (1992–1993), p.13.
44. *O.tidene* 25.5.1987, sak 2 nr 2.
45. The Norwegian Parliament (Stortinget) consists of 169 representatives. Until 2009, when passing proposed laws, Parliament would divide into two chambers: the Odelstinget and Lagtinget (where the latter consisted of one fourth of the representatives). Proposed laws were first debated and voted on in the Odelstinget and, if passed, then sanctioned (or not) by the Lagtinget. Today, proposed laws are debated in a plenary session. The voting record I cite is from the debate in the then Odelstinget. Each proposal was voted on separately, and the law *in toto* at the end. The entire law was passed with 67 for and 27 against; the principle of anonymity: 53 for, 42 against; treatment only to be offered to married couples, 49 for, 45 for opening treatment for cohabiting couples. See *O.tidene* 25.5.1987, sak 2.
46. The public health-care system in Norway is central to the welfare state. Public health-care services are designed to be equally accessible to all residents, independent of social status. The services are financed through taxes. Private health-care services have to be authorized. Health issues are also debated in terms of the priority they are to be given within the health-care system.
47. See also Sirnes (1997: 221). This 'deal' was one that several of those I interviewed also commented on. See also *O.tidene* 25.5.1987, sak 2 nr 2.
48. Innst. O. nr.60 (1986–1987), p.11.
49. See the Norwegian Official Report: NOU (1987: 23). This report suggested that involuntary childlessness be given a very low priority. Two arguments are relevant to the Norwegian position: one has to do with the medicalization of infertility (whether the condition of being involuntary childless is to be considered an illness or 'fate'); the other has to do with the debate about private versus public health-care services. The Conservative Party was for giving involuntary childlessness a low priority and was also for letting private clinics offer IVF treatment, something the Labour Party would not accept. See Sirnes (1997: 265–70) for further discussion of this issue and how infertility treatment is legitimized in Norway and the UK.
50. NOU (1991: 6); see also RMF (1983).

51. St.meld. nr.25 (1992–1993).
52. St.meld. nr.25 (1992–1993), presented by the Norwegian government to Parliament on 12 March 1993. An abbreviated English version of this report was issued by the Ministry of Health and Social Affairs: 'Biotechnology Related to Human Beings', Report No.25 (1992–1993), available at: http: //www.helsetilsynet.no/htil/avd2/bio_act.htm, retrieved 26 October 1999. The White Paper was in turn based on an earlier report submitted by the Ethics Committee: NOU (1991: 6). Overall, the White Paper tended to be more liberal than the report from the Ethics Committee.
53. Lov (1994–08–05, nr 56), hereafter the Biotechnology Act.
54. The translation I use is taken from the official English version of the Act. However, the literal translation of the original Norwegian wording would be 'in a society where there is room for all' – the connotation is slightly different.
55. Innst. O. nr.60 (1986–1987), p.3.
56. See Innst. O. nr.67 (1993–1994); Ot.prp. nr 37, p.11. The issue of egg donation had been debated in Parliament in conjunction with the earlier White Paper (St.meld. nr.25, 1992–1993), and it was already then obvious that there was no majority for this proposal.
57. There were two major reasons for importing sperm from Denmark. One had to do with the difficulty of getting enough Norwegian men to donate sperm. This situation was exacerbated when the need to screen sperm for HIV and other infectious diseases (involving a quarantine period of six months) became paramount. The other reason was capacity and costs: the sperm bank in Denmark was efficient. With the transition to the use of frozen sperm, it was on the whole simpler and more cost effective to import from Denmark (see Melhuus 2003).
58. Out of a total of twenty-five ministers, five were Christian Democrats (in addition to the prime minister), including those of Culture and Church Affairs, International Development, and Children and Family. All of these, significantly, reflect the party's main concern with 'values': the family, the Church, and international aid.
59. Innst. O. nr.67 (1993–1994), p.21.
60. St.meld. nr.14 (2001–2002).
61. Meeting in Parliament, Monday 17 June 2002; item 4. An interesting observation made by the chairperson, representative Ballo, in his introduction is the fact that the political constellation is more or less the same as in 1994: the political parties have more or less the same view on both occasions.
62. For the revised Act, see Lov (2003–12–05, nr 100). For the record of votes, see: http: //www.stortinget.no/no/Saker-og-publikasjoner/Publikasjoner/Referater/Odelstinget/2003-2004/031118/voteringer/
63. Meeting in Odelstinget, Tuesday 18 November 2003 at 12.10 pm.
64. Innst. S. nr.238 (2001–2002), p.8. The Norwegian text reads: *Etter flertallets syn skal det være en lovmessig forutsetning for den assisterte befruktningen at det biologiske opphavet klart kan defineres som én biologisk mor og én biologisk far.*
65. Also significant was the fact that Sweden had passed a law in 1985 prescribing the use of known sperm donors.
66. For a discussion of the adoption laws and laws related to assisted conception, see Melhuus and Howell (2009); see also Meeting in Parliament, Monday 17 June 2002, item 4 and St. meld. nr.14 (2001–2002).
67. Innst. S. nr.238 (2001–2002), p.10.
68. For a polemical argument, see Hazekamp (2005).
69. It is therefore interesting to note that in 2011 the Norwegian Broadcasting Corporation (NRK) reported that there was a scarcity of donor sperm in Norway, and that the two major fertility clinics offering sperm donation were not accepting new couples (see: http: //www.nrk.no/nyheter/1.7519093, retrieved 15 March 2011). The increased demand for sperm must be seen in conjunction with the Marriage Act of 2008 (Lov 2008–12–19, nr 112), which among other things granted lesbian couples the right to infertility treatment (see Chapter 4).
70. Lov (2003–12–05, nr 100), §§ 2.5–2.12.

71. Lov (2003–12–05, nr 100), §§ 2.16 and 2.17.
72. Lov (2003–12–05, nr 100), § 2–5.
73. This is suggested by Chairperson Ballo in his introductory remarks. Meeting in Parliament, Monday 17 June 2002, item 4.
74. See the introduction (2 *Innledning*) to the bill to Parliament, where a special section is dedicated to the precautionary principle: Ot. prp. nr.64 (2002–2003), 3–4.
75. See Lov (2007–06–15, nr 31). See also Meeting in Odelstinget, 24 May 2007; the discussion in Parliament was based on two documents: Innst. O. nr.62 (2006–2007) and Ot. prp. nr.26 (2006–2007).
76. Forhandlinger i Odelstinget nr 29, 24 May 2007, pp. 401–2.All the ensuing quotes are taken from Meeting in Odelstinget nr 29, 24 May 2007.

Chapter 4

1. *Verdens Gang* is a major Norwegian tabloid newspaper. The article draws on the Save the Children report 'Women at the Front Lines of Health Care: State of the World's Mothers' (2010). For the original article, see: http: //www.tv2nyhetene.no/innenriks/norge-beste-moedreland-i-verden-3199753.html, retrieved 17 June 2010.
2. Act nr 2008 – 06 – 27, nr 53. It is important to stress that legislation providing for same-sex marriage is – on a global scale – indeed exceptional, indicating the culturally specific forms of arguments tied to this form of family formation.
3. Lov 1997–06–13, nr 39, chapter 1A, § 2. This provision excludes surrogacy arrangements as a legal practice in Norway. Nevertheless, the Act explicitly specifies that 'An agreement to give birth to a child for another woman is not binding'. Note that to give birth for a man is not even mentioned.
4. I am here disregarding the possibility of cloning, but see Stolcke (2009). For a gendered analysis of the symbolic representations of egg and sperm, see Martin (1991).
5. But see Daniels (1998: 96), who argues that there is a qualitative difference between sperm and blood
6. This is also a point Daniels makes: that semen was considered independent of the men who donated it. Weight was placed on the quality of sperm, and not on the psycho-social qualities of donors (Daniels 1998).
7. Countries that have rescinded the anonymity clause are Sweden (1984), Norway (2003), the Netherlands (2004), Great Britain (2005), Finland (2006) and Belgium (2007). See *Le Monde*, 7 May 2009.
8. Ot. prp. nr.25 (1986–1987), p 15, my translation.
9. Ot. prp. nr.25 (1986–1987), p.15.
10. Ot. prp. nr.25 (1986–1987), p.32.
11. Ot. prp. nr.25 (1986–1987), p.32.
12. See Tranøy (1989) for a discussion of reproductive technologies and rights.
13. Ot. prp. nr.25 (1986–1987), p.33.
14. O.tidene 25.5 Sak nr 2; Stortingsforhandlinger sesjon 1986–87, pp. 308–45.
15. See Lov 1981–04–08, nr 07, chapter 2 § 6 (who is the father of the child); Ot.prp nr 93 (2001–2002); Lov om endringer i lov av 8. April 1981, nr 7 concerning ascertaining a change in paternity. This amendment granted the child, each of the parents and a man who claims he is father to the child to have the case tried in court.
16. A woman who has given birth is obliged to state who the father of the child is; if she does not, the state has the responsibility to determine the child's father (Lov 1981–04–08, nr 07, chapter 2 § 5).
17. All quotes in this paragraph are taken from Ot. prp. nr.93 (2001–2002), § 4.2.2; translations are mine. Several of those heard in this process express concern about the insecurity and

unrest that suggested amendments in the law might imply for the children concerned. Yet the argument about the significance of knowing one's biological origin has the most adherents.
18. For a discussion of origins in connection with donor insemination (DI) children and parallels to adoption, see Haimes (1998: 58).
19. The French state, by contrast, in suggesting amendments to its bioethics law, which include a proposal to rescind the anonymity clause (for both egg and sperm donation), states explicitly in an article in *Le Monde* that this is not a genetic argument but rather that donor anonymity 'deprives the child of a dimension of their history' (Chemin 2009). The journalist quotes the Conseil d'Etat: 'Ce principe édicté en 1994 comporte à long termes des effets préjudiciables à l'enfant, essentiellement parce que ce dernier est privé d'une dimension de son histoire'. And she continues: 'Contrairement, ce que l'on dit souvent, la levée de l'anonymat n'est pas fondée sur une approche "génétique" de l'ascendance: la recherche du donneur est, note le Conseil d'Etat "une démarche tendant à mieux se construire personnellement et psychologiquement, non dans le but d'avoir une autre mère, mais pour ne pas vivre dans l'ignorance ou même le mensonge"' (Chemin 2009). In other words, living a lie is also an argument.
20. Ot. prp. nr.25 (1986–1987). In order to weight its arguments in formulating the proposed legislation, the document refers to persons/institutions heard. Translations are mine.
21. The Norwegian term *befruktning* can translate either as conception or fertilization. Another Norwegian term for conception is *unnfangelse* but it is seldom used in conjunction with these debates. In my translation I use conception and fertilization interchangeably.
22. Ot. prp. nr.25 (1986–1987), p. 19.
23. Stortingsforhandlinger 1988–89 nr 41, p. 4004 (the debate in Parliament on biotechnology, 1989). At that time maternity was not yet legally defined.
24. Quotes in the remainder of this paragraph are taken from parliamentary debates in 1993 and are stated by Magnar Sortåsløkken (male, Socialist Left Party), Ole Johs Brunæs (male, Conservative Party), and Kirsti Kolle Grøndahl (female, Labour Party). See Stortingstidende 10.6.1993: sak 1. *Stortingsforhandlinger sesjon 1992–93*, pp. 4346–87; all translations are mine.
25. For a brief presentation of anonymous birth in France, see Lefaucheur (2004).
26. However, the point about the right to know one's biological origin was a contentious issue in the formulation of the Hague Convention on Inter-country Adoption (1993). Whereas nearly all countries receiving children were in favour of including such a clause, most donor countries disagreed and such a clause was not included (Melhuus and Howell 2009). See also Dolgin (1997) for a discussion of adoption laws and their impact in the USA.
27. Lov (1981–04–08, nr 07), chapter 2, § 5.
28. The seminar was an open meeting about reproductive tourism held in conjunction with a small temporary exhibit entitled 'The Cyber Stork'. There were nine speakers, among them the leader of the FUB's successor organization, Ønskebarn; a philosopher; heads of fertility clinics in Norway and abroad; and a homosexual parent (Odd Jenvin). I had also been asked to give a short presentation. For details of the case discussed below I draw on Jenvin's presentation at the seminar and our subsequent correspondence. The program and presentations have been gathered and made available online (see: http: //www.bion.no/publikasjoner.shtml); all translations are mine. See also Solerød (2008).
29. More recently, India is becoming an alternative for those seeking surrogacy arrangements. Surrogacy services in India are less expensive than the USA. Questions have been raised concerning ethical aspects of these services, pointing to the possible exploitation of poor women, as well as the legal aspects.
30. See Sandberg (2009) for a discussion of the principles of equality and how they pertain to the legal status of the child.
31. If you recall, with the revised Marriage Act (Lov 2008–12–19, nr 112), same-sex couples are granted the same rights as heterosexual couples, and lesbian couples are granted the right to infertility treatment.
32. For different discussions of surrogacy practices and notions of motherhood, see Dalton (2000), Ragoné (1994) and Kahn (2002).

33. This is a complicated case, but turns on the significance of the biological principle, while indicating how citizenship, maternity, and adoption processes come together. The case had high media coverage and was also presented in a documentary series by the Norwegian Broadcasting Corporation (NRK) called *Brennpunkt* (transmitted 22 March 2011). Various public persons have called for a resolution, arguing that the best interests of the children must take priority over other concerns. See also Kroløkke (2012) for a discussion of citizenship related to transnational surrogacy in India. She uses the Volden case as her example.
34. In fact there was a case (in 2007) where a woman in the course of IVF treatment contracted cancer and had to remove her uterus. She applied to take her fertilized eggs abroad in order to make use of a surrogate. The then Minister of Health (Sylvis Brustad) granted permission – and was then accused (in the media) of breaking Norwegian law. See *Vårt Land*, 16 June 2007; *Dagsavisen*, 16 and 22 June 2007; *Aftenposten*, 19 and 20 June 2007.
35. See Riksaasen (2001) for a discussion of lesbian couples in Norway and their procreative practices.

Chapter 5

1. http://web.retriever-info.com/services/webdocument.html?documentId=002411200507013 1759398&serviceId=2, retrieved 17 March 2011.
2. See Lov (1994–08–05, nr 56).
3. For example, in cases where the Norwegian state bases some of its medical practices on research carried out elsewhere but not permitted in Norway.
4. Rose does not use the terms 'ours' and 'we' unproblematically but is basically referring to what he calls advanced liberal democracies.
5. Rose phrases it thus: 'once each life has a value that may be calculated, and some lives have less value than others, such a politics has the obligation to exercise this judgement in the name of the race or nation' (Rose 2007: 57).
6. See Wade (2007) for a different approach to the mutually imbricated issues of race, nation, kinship and genetics; and see especially Campbell's contribution comparing Spain, Norway and the UK with regard to gamete-matching regulation and immigrant integration policies (Campbell 2007). Campbell states: 'The contexts in which genetic knowledge is becoming meaningful in Europe are located at a time when ideas of national identity, consumption practices and bodily values are undergoing rapid transformation ... This calls for a recognition that the dynamics and contexts of contemporary racializations are fundamentally different from early twentieth century eugenics' (Campbell 2007: 116).
7. See Cohen (1999) for an elaboration of the notion of ethical publicity.
8. Kvande (2008: 133) writes that the term 'sorting society' was first used in 1982 in connection with the debates in Parliament on the application of amniocentesis based on St.meld. nr.73 (1981–1982).
9. Becker, in contrast to Rose, suggests that in the wake of commercialization 'a new form of eugenics is emerging' (Becker 2000: 245).
10. I must add that this legislation also proposed the continued prohibition on research on embryos. In the debate in Parliament, the term sorting society was also used in this connection. The problem was posed as an opposition between prohibiting research on embryos while at the same time allowing for the destruction of supernumerary embryos. For example, the Labour Party representative Asmund Kristoffersen says: 'Research on fertilized eggs is of vital significance. When the government assumes (*legger til grunn*) that an individual's integrity starts at conception, the result is a restrictive attitude to assisted conception and prenatal diagnosis. At the same time it is accepted that supernumerary fertilized eggs ... are to be destroyed. This must at least be a dilemma. Research on supernumerary fertilized eggs could provide some useful answers for possible treatments for sick people' (Meeting, Odelstinget, Tuesday 18 November 2003).

11. Quotes from the transcript of Høybråten's speech are taken from the website of the Norwegian ministries: http: //odin.dep.no/hod/norsk/aktuelt/taler/minister_a/042071-990284/dok-bn. html, retrieved 2 December 2009. The speech can now also be downloaded from: http: // www.regjeringen.no/en/archive/Bondeviks-2nd-Government/hd/265625/266491/ny_lov_ om_bioteknologi.html?id=266994. All translations are mine.
12. Lov (1994–08–05, nr 56), § 1.1.
13. Much of the debate in Parliament revolved around the application of ultrasound and what ultrasound should be used for. This debate can be traced to earlier heated discussions about the applications of ultrasound in Norway; see Kvande (2008) for an excellent documentation of this history. Ultrasound was first offered as a routine control at one of the major hospitals in Oslo (Ullevål sykehus) in 1981, provoking strong reactions. One of the main controversies about the use of ultrasound concerns the question of prenatal diagnosis (Kvande 2008: 116). The debate turns on the blurring of the boundaries between pregnancy care and fetal care, and the creation of the fetal patient.
14. For a brief historic overview of abortion based on eugenic indication in Norway, see Giæver (2005); see also Nielsen, Monsen and Tennøe (2000: 195–97).
15. It is interesting to note that neither Høybråten (nor his supporters) use specific Christian arguments to ground their views. This is in contrast to the British parliamentary debate on embryo research in the 1980s. Mulkay (1997: esp. 96–115) shows how these debates were phrased in terms of religion versus science, faith versus reason.
16. The following quotes are taken from the debate in Odelstinget, Meeting, 18 November 2003.
17. Solberg bases his statement on a report from the National Health Institute on late abortions that have to be approved by a special committee (Mo et al. 2006). According to the report, the number of such abortions in 2005 was 45. Recent figures released from the Medical Birth Register (*Medisinsk fødselsregister*) confirms that about 90 per cent of pregnant women who have had PND and a fetus detected with Down syndrome choose to abort. This number has been stable over the last ten years (Skodje 2011).
18. The term 'screening' to denote routine ultrasound examinations is not generally used in Norway. Kvande suggests that this may have to do with the negative connotations of screening, associated with state eugenic control, and hence also with the aversion to a sorting society (Kvande 2008: 289).
19. For a different view, see Lavik (1998). In his book about the intellectual roots of racism, Lavik argues that scientists were among the foremost of those to establish the premises for racism (Lavik 1998: 12). In particular he mentions genetics, medicine and psychiatry, the latter being his main concern. Thus he counters arguments that racism is primarily lodged in lay opinions and argues that racist ideas at the time (in Norway as elsewhere) must be understood within an international context. He states that with regard to the implementation of sterilization laws in Norway, psychiatrists played a significant role (Lavik 1998: 39; see also Nielsen, Monsen and Tennøe 2000: 95–117). For a more detailed discussion of mental hygiene (which also included sexual hygiene) in Norway, see Seip (1994: 119–23).
20. For a detailed analysis of the development of Norwegian welfare policies from 1920 to 1975, see Seip (1994).
21. See Lov (1934–06–01, nr 2). However, sterilization was not the only field of social policy where eugenic considerations played a role. The Norwegian Marriage Act of 1918 contained a clause prohibiting marriage for the insane (Lov 1918–06–31). This law was not changed until 1991 (Broberg and Roll-Hansen 2005: xii). For a discussion of eugenic indicators in relation to abortion, see Giævær (2005).
22. According to Haave, the health authorities received 47,000 applications for sterilization between 1934 and 1977. Of these about 44,000 were carried out, 30,000 being done so on women. Although most of these interventions were done on the basis of a demand from the woman herself, it is difficult to ascertain whether these were truly voluntary sterilizations (St. meld. nr.44, 2003–2004, § 15.2.1).

23. The first abortion debates in Norway were raised in the wake of a newspaper article written by the feminist Katti Anker Møller in 1913. These early debates revolved around the use of social indicators as grounds for abortion; eugenic indicators were not even mentioned (Giæver 2005: 3472).
24. In addition to the women's movement and its struggle for procreative control, there is the development of psychiatry and its increasing influence, especially within correctional services and criminal law (the debates on social and mental hygiene), as well as the role played by young medical doctors, the Church and so on (see Nielsen, Monsen and Tennøe 2000: 88–123).
25. Lavik, albeit from a different perspective, also points to the alliance between scientific expertise and political authorities and the implications of this alliance for power relations (Lavik 1998: 26).
26. See St.meld. nr.44 (2003–2004).
27. As Syse states: 'The law had three indications for abortion: medical, eugenic and ethical. The eugenic indication (regulated in § 1, nr.2 [of Lov 1960–11–11, nr 72]) was based on the condition that there was a serious risk that the "child may come to have a serious illness or a major physical or mental (*sjelelig*) defect"' (Syse 1993: 25, my translation; see also Giæver 2005: 3474).
28. The Norwegian title of this report uses the word *tatere*, the now discredited Norwegian term used to designate travelling people.
29. St.meld. nr.15 (2000–2001), quoted in St.meld. nr.44 (2003–2004), § 31.1.
30. The Mission for the Homeless was a private organization, originally founded in 1867 by the priest Jacob Walnum and then known as the Foreningen til motarbeidelse av omstreifervesent; it changed name to the Mission for the Homeless in 1935. It was run by priests ordained in the Norwegian Lutheran Church, received public funding, and was given the responsibility of working with Romany people. It was closed down in the late 1980s (see: http: //www.kildenett. no/ordbok/1223459877.68, retrieved 15 December 2009; and http: //snl.no/romanifolk, retrieved 8 December 2009). According to a government White Paper, the Mission saw sterilization as an important component in solving the problem of vagabonds (*omstreiferproblemet*). Sterilization became an explicit dimension of the Mission's activities from the middle of the 1930s until the end of the 1940s (St.meld. nr.44, 2003–2004, § 15.2.2).
31. See St.meld. nr.44 (2003–2004). The document quotes Haave, stating: 'There are indications that more than 230 *tater* women were sterilized outside of the legal framework. If we add the number of interventions carried out within the terms of the law (*med hjemmel i lov*), this implies that over 300 *tater* women may have been sterilized from the 1930s to the 1970s' (my translation).
32. According to an article in *Store norske leksikon*, one important aim of the Mission (in addition to its missionary work) was to enforce permanent settlement among travellers. They established several orphanages where children from Romany families were forcefully placed. In the period between 1900 and 1947, an estimated 1,000 children were resettled either in orphanages or in foster families (see: http: //www.snl.no/romanifolk, retrieved 8 December 2009).
33. 'Kirkemøtet 2000 ber taterne om unnskyldning og tilgivelse for den urett og de overgrep som fra kirkens side er begått mot deres folk', http: //kirken.no/index.cfm?event=doLink&famId=5080, retrieved 6 January 2010. See also http: //www.snl.no/romanifolk, retrieved 8 Decemeber 2009.
34. Aasen (1990), reflecting on the links between, on the one hand, racial hygiene and the sterilization laws and, on the other, the regulation of PND and selective abortion, points to two aspects that reinforce this link: that PND is spoken of as 'preventive health care' and that PND raises the same questions as did racial hygiene. What lives have value and are worth living?
35. The involvement of the Church can be traced through Church documents, which also reflect the nuances in the debates. See, e.g., documents from the Church Meeting (*Kirkemøte, KM*): 1985: KM 3/85 Menneskesyn og Menneskeverd; 1987: KM 20/87 Kirkens engasjement for det ufødte liv; 1989 KM 12/89 Bioteknologi og menneskeverdet; 13/89 Abortlovgivningen. See

also the *Kirkerådet* 1989; Bishops Meeting 1989, 16/89 'Etiske spørsmål vedrørende det ufødte liv'. Church opinions are also to be read in various hearing documents accompanying legislative processes.
36. See Kooten Niekerk (1994); and for a more recent discussion on the moral status of the embryo, see Østnor (2008). In conjunction with the parliamentary debate on embryo research in the UK, Mulkay shows that there was no single distinctive Christian view. Rather, he argues that the fragmentation of religious opinions seriously weakened the position of Christians in these debates. (Mulkay 1997: 110; see also Franklin 1999).
37. A seminar on 'Religion and Bioethics' was organized in 2005 by Jan Helge Solbakk (Professor and Head of the Section for Medical Ethics, University of Oslo). Lecturers included Laurie Zoloth (Professor of Medical Ethics, Northwestern University, USA), Donald Bruce (Director of the Church of Scotland's Society, Religion and Technology Programme) and Dr Aida Ibrahim M. Al Aqeel (Senior Scientist at King Faisal Specialist Hospital and Research Centre, Riyadh, Saudi Arabia). A 2005/6 research project 'The Moral Status of Human Embryos with Special Regard to Stem Cell Research and Therapy' was financed by the Norwegian Research Council. See Baune et al. 2007.
38. Keep in mind that in Norway the pre-embryo (as an entity) has not been recognized; this in contrast to the UK (see Sirnes 1997).
39. This is a complex issue that deserves more attention than I can give it here. However, it is important to keep in mind that there is no consensus on this matter either within Christianity itself (e.g., between Catholics and Protestants) nor within the Norwegian Church.
40 Lov (1994–08–05, nr 56), § 1.1.
41. This distinction is also reflected in gradual shifts in the meaning and application of ultrasound. Until the mid 1980s ultrasound was used primarily as a technique in pregnancy control; thereafter it became a technique primarily used in prenatal diagnosis. This represented a shift in focus from the pregnant woman to the fetus (Kvande 2008: 284). Pertinent here is the notion of 'mother-belonging' (*morstilhørigheten*) mobilized to challenge egg donation and that the right to self-determined abortion is based on an idea that the embryo/fetus is an integral part of the woman's body, and not a separate entity.
42. Aasen (1990) in her discussion of PND intimates that it is perhaps wrong that society gives individual women this choice.
43. Again, this is in stark contrast to the debates on the sterilization laws, where economic arguments were explicitly applied as part of the justification: sterilization would save the state money (Roll-Hansen 2005: 170).
44. See Christoffersen (1986) for a theological cum ethical reflection about the unresolved problems of abortion in conjunction with biotechnology. See also Solbakk (2005) for critical reflections on ethics and Norwegian biolegislation.
45. See Lov (2007–06–15, nr 31). Again I refer to Sirnes (1997) comprehensive comparative study of the parliamentary debates in the UK and Norway (with regard to biotechnology and research on embryos). One of the important points he demonstrates is how the ontological question of the moral status of the embryo never gained much ground in Norway, whereas in the British parliamentary debates this was at the heart of the matter.

Chapter 6

1. However, see Donchin (2010) on the stabilizing effect of reproductive travel in countries with restrictive policies.
2. To give two other examples: The Norwegian law defining death was changed in 1977 making so-called 'brain-death' the definitive criteria (see For 1977–06–10, nr.02). This redefinition of death, as we know, allows for organ transplantation (see Lock 2002, where she provides an excellent comparative study of death and its meanings in the USA and Japan). In 1979 Norway amended its 1964 law regarding personal names (Lov 1964–05–29, nr 1; Lov 1979–

06–08, nr 39). Until then surnames in Norway followed the agnatic principle, whereby a legally married wife and her children automatically took her husband's surname. With this amendment, a child has to be registered with a first name and a surname within six months of birth. The surname of the mother or father was equally acceptable. If no registration was made, the child would be given its mother's surname. This implied a shift from a predominantly patrilineal system to a matrilineal one, in so far as the woman had not changed her surname to that of her husband.
3. See NOU (2009: 5). There is an English summary of the report on pages 16–18. The report became available only when I had concluded my research, but its significance meant that I could not ignore it. Significantly, the committee goes under the name *farskaps utvalget* – that is, 'the paternity committee'.
4. NOU (2009: 5), p.16; all translations from the report are my own.
5. See Lov (2008–12–19, nr 112).
6. NOU (2009: 5), p.19.
7. NOU (2009: 5), pp.32, 33.
8. The committee also suggests upholding the rule of *pater est* for married couples but not extending the rule to cohabiting couples (NOU 2009: 5, p.78).
9. NOU (2009: 5), p.31
10. NOU (2009: 5), p.83.
11. In other words, if maternity has been transferred in any other way, it is not recognized by Norwegian authorities (NOU 2009: 5, p.88).
12. The report discusses the different cases/groups that might make use of a surrogate (NOU 2009: 5, pp.84–95). I will however, only consider the male same-sex couple. The report also discusses lesbian married couples who make use of anonymous donor sperm contrary to Norwegian legislation, which only attributes co-motherhood to the woman in the couple who has not given birth when the sperm comes from a known donor (NOU 2009: 5, pp.91–93). They suggest that these situations be treated analogously with heterosexual couples that have had a child with anonymous donor sperm. In the latter case the legislation does not demand that the donor be known in order to grant fatherhood. Therefore, it would imply differential treatment of children born of lesbian and heterosexual couples (in other words an equality principle is applied). In cases where anonymous sperm has been used (at clinics abroad), the committee proposes that co-motherhood status be granted to the woman who has not given birth. In other words, she should not have to adopt the child.
13. The committee finds it very unlikely that a person or couple who has engaged a surrogate mother not know her name and address, and that the surrogate mother must be known to the biological father. Withholding the surrogate's name is done by volition. Thus it is possible to obtain her identity and hence her consent. This is in contrast to couples who have made use of an anonymous sperm donor, where the donor's identity is unknown (to the mother) and the donor cannot (according to the law) be ascribed paternity. Therefore his consent is not necessary for adoption procedures.
14. NOU (2009: 5), p.94.
15. NOU (2009: 5), pp.88–90.
16. NOU (2009: 5), p.95.
17. NOU (2009: 5), p.91.
18. For example, the committee states that in a situation where knowledge of genetic origin is pitched against establishing stable parenthood, the former must yield (NOU 2009: 5, pp.91–93), and this despite legislators' wish that most children should have this right. This is stated in conjunction with the practice of AID abroad and the demand on the part of Norwegian legislators that known donors be used in order that co-motherhood can be established.

Postscript

1. Kari Anne Ulfnes assisted me in carrying out these (and some other) interviews. Sometimes we did these together, some interviews she did on her own.
2. However, I would nevertheless argue that the involuntary childless do form a community of sorts, if primarily in abstract and imaginative terms. They have, in a sense, been framed, especially by the media, but also by acts of legislation. Moreover, they are present as an interest group through the many activities of their association.
3. It is perhaps important to keep in mind that since I began this research over ten years ago the issues pertaining to infertility and being involuntarily childless have become much more public – and hence perhaps also more legitimate. Nevertheless, the fact of greater legitimacy does not deter from the sensitive and emotional qualities of the experience of those concerned.

Appendix

1. The National Bureau of Statistics' figures are available at: http://www.ssb.no/fodte/. Retrieved 16 February 2010.
2. NOU (2008: 18), § 4.2.
3. See Frønes, Jensen and Solberg (1990) for an overview of childhood in Norway.
4. In July 2009 parents were given the combined right to 56 weeks of paid parental leave, 10 of which must be taken by the father.
5. See the Cash Support Act (Lov 1998–06–26, nr.41). In 2009 the amount given was Nkr 3,303 per month per child (1 Euro is approximately 8 Nkr.)

REFERENCES

Explanatory note

The Norwegian alphabet consists of 29 letters. The last three letters are 'æ', 'ø' and 'å' (also written as 'aa'). Hence, surnames beginning with 'aa' (such as Aasen) will be listed at the end of the reference list, as will those beginning with 'ø' (such as Østnor).

Norwegian government documents are referenced following the standard abbreviations. Below follows a list of abbreviations. Legal acts are referred to as 'lov'. English translations of the titles of acts are provided.

Stortinget is the Norwegian term for the Norwegian Parliament.

Odelstinget refers to the largest of the two chambers in Parliament (see Chapter 3, footnote 45).

Innst.	Innstilling (Recommendation)
Innst. O.	Innstilling til Odelstinget (Recommedation to Odelstinget)
Innst. S.	Innstilling til Stortinget (Recommendation to Stortinget)
NOU	Norges Offentlige Utredninger (Norwegian Official Reports)
Ot.prp.	Odelstingsproposisjon (Proposition to the Odelstinget)
O.tidende	Odelstingstidende (the Official Report on Odelsting Proceedings)
St.prp.	Stortingsproposisjon (Proposition to the Stortinget)
Stortingsforhandlinger	the Official Record of Storting Business
Stortingstidende	the Official Report on Storting Proceedings (comparable to Hansard (UK) and Congressional Record (USA).
St.melding	Stortingsmelding (Official Report to Stortinget, White Paper)

Norwegian Government Documents

Recommendations to Parliament
(Innstillinger til Stortinget/Odelstinget)

Innst. 1953. 'Innstilling fra inseminasjonskomiten'.
Innst. O. nr.60. 1986–1987. 'Innstilling fra sosialkomiten om lov om kunstig befruktning'.
Innst. O. nr.67. 1993–1994. 'Innstilling fra sosialkomiten om lov om medisinsk bruk av bioteknologi (Ot. prp. nr.37)'.
Innst. S. nr.238. 2001–2002. 'Innstilling fra sosialkomiten om evaluering av lov om medisinsk bruk av bioteknologi, St.meld. nr.14 (2001–2002)'.
Innst. O. nr.16. 2003–2004. 'Innstilling fra sosialkomiteen om lov om humanmedisinsk bruk av bioteknologi m.m. (bioteknologiloven)'.
Innst. O. nr.62. 2006–2007. 'Innstilling fra helse–og omsorgskomiteen om lov om endringer i bioteknologiloven (preimplantasjonsdiagnostikk og forskning på overtallige befruktede egg)'.

Legal Acts

FOR 1977–06–10, nr 02. 'Forskrifter om dødsdefinisjon i relasjon til lov om transplantasjon, sykehusobduksjon og avgivelse av lik mm' [Regulation Regarding the Definition of Death in Relation to the Law on Transplantation, Hospital Autopsy and Relinquishing of Corpses].
Lov 1917–04–02, nr 1. 'Adopsjonsloven: At knytte et baand' [Adoption Act: To Forge a Link].
Lov 1918–05–31, nr 2 'Om ingaaelse og oppløsning av egteskab' [Act Relating to the Entering and Dissolution of Marriage].
Lov 1934–06–01, nr 2. 'Lov om steriliseirng (steriliseringsloven)' [Sterilization Act].
Lov 1960–11–11, nr 72. 'Lov om svangerskapsavbrudd' [Abortion Act].
Lov 1964–05–29, nr 1. 'Lov om personnavn' (navneloven) [Act Relating to Personal Names].
Lov 1975–06–13, nr 50. 'Lov om svangerskapsavbrudd (abortloven)' [Abortion Act].
Lov 1979–06–08, nr 39. 'Lov om personnavn' [Act Relating to Personal Names].
Lov 1981–04–08, nr 07. 'Lov om barn og foreldre (barneloven)' [Children's Act].
Lov 1987–06–12, nr 68. 'Lov om kunstig befruktning' [Artificial Procreation Act].
Lov 1991–07–04, nr 47. 'Lov om ekteskap (ekteskaploven)' [Marriage Act].
Lov 1993–04–02, nr 38. 'Lov om framstilling og bruk av genmodifiserte organismer mm (genteknologiloven)' [Gene Technology Act].
Lov 1994–08–05, nr 56. 'Om medisinsk bruk av bioteknologi (bioteknologiloven)' [Act Relating to the Application of Biotechnology in Medicine/Biotechnology Act]
Lov 1997–06–13, nr 39. Lov om endringer i lov av 8. April 1981, nr 7 [Amendments to the Children's Act].
Lov 1998–06–26, nr 41. 'Lov om kontantstøtte til småbarnsforeldre (kontantstøtteloven)' [Act Relating to Cash Support].
Lov 2002–06–07, nr 19. 'Lov om personnavn (navneloven)' [Act Relating to Personal Names].

Lov 2003-12-05, nr 100. 'Lov om humanmedisinsk bruk av bioteknologi m.m. (bioteknologiloven)' [Act Relating to the Application of Biotechnology in Medicine, etc./Biotechnology Act].

Lov 2007-06-15, nr 31. 'Lov om endringer i bioteknologiloven (preimplantasjonsdiagnostikk og forskning på overtallige befruktede egg)' [Act Relating to Changes in the Biotechnology Law (Preimplantation Diagnosis and Research on Supernumerary Embryos)].

Lov 2008-06-27, nr 53. 'Lov om endringer i ektekskapsloven, barneloven, adospjonsloven, bioteknologiloven mv (felles ekteskapslov for heterofile og homofile)' [Act Relating to Changes in the Marriage Law, Child Law, Adoption Law, Biotechnology Law etc.].

Lov 2008-12-19, nr 112. 'Ekteskapsloven' [Marriage Act].

Norwegian Official Reports (Norges Offentlige Utredninger)

NOU 1977: 35. 'Lov om barn og foreldre (barneloven)'.
NOU 1987: 23. 'Retningslinjer for prioritering innen norsk helsetejeneste'.
NOU 1991: 6. 'Mennesker og bioteknologi'.
NOU 1997: 18. "Prioritering på ny: Gjennomgang av retningslinjer for prioriteringer innen norsk helsetjeneste'.
NOU 2008: 18. 'Fagopplæring for fremtida'.
NOU 2009: 5. 'Farskap og annen morskap: Fastsettelse og endring av foreldreskap'.

Helsedirektoratet Reports
Helsedirektoratet (Directorate of Health). 2011. 'Evaluering av bioteknologiloven: Status og utvikling på fagområdene som reguleres av loven'.

Parliamentary Propositions (Odelstings proposisjoner)

Ot. prp. nr.25. 1986–1987. 'Om lov om kunstig befruktning', Sosialdepartementet.
Ot. prp. nr.37. 1993–1994. 'Om lov om medisinsk bruk av bioteknologi', Sosial–og helsedepartementet.
Ot. prp. nr.27. 1996–1997. 'Om lov om endringer i lov om medisinsk bruk av biotknologi (forbud mot genetisk testing av kjønnstilhørighet for annet enn medisinske formål)'.
Ot. prp. nr.21. 1997–1998. 'Om lov om endring i lov om medisinsk bruk av bioteknologi (forbud mot fremstilling av arvemessig like individer)'.
Ot. prp. nr.93. 1998–1999. 'Om lov om endringer i lov 5 august 1994 nr.56 om medisinsk bruk av bioteknologi', Helse–og omsorgsdepartementet.
Ot. prp. nr.93. 2001–2002. 'Om lov om endringer i lov 8 april 1981 nr.07 om barn og foreldre (fastsettelse og endring av farskap)'.
Ot. prp. nr.64. 2002–2003. 'Om lov om medisinsk bruk av bioteknologi m.m. (bioteknologiloven)', Helse–og omsorgsdepartementet.
Ot. prp. nr.26. 2006–2007. 'Om lov om endringer i bioteknologiloven (preimplantasjonsdiagnostikk og forskning på overtallige befruktede egg)'.

Official Reports to Parliament – White Papers (Stortingsmeldinger)

St.meld. nr.73. 1981–1982. 'Om organisering av medisinsk–genetiske servicefunskjoner i Norge'.
St.meld. nr.8. 1990–1991. 'Om bioteknologi'.
St.meld. nr.25. 1992–1993. 'Om mennesker og bioteknologi'.
St.meld. nr.14. 2001–2002. 'Evaluering av lov om medisinsk bruk av bioteknologi'.
St.meld. nr.15. 2000–2001. 'Nasjonale minoritetar i Noreg: Om statleg politikk overfor jødar, kvener, rom, romanifolket og skogfinnar'.
St.meld. nr.29. 2002–2003. 'Om familien – forpliktende samliv og foreldreskap'.
St.meld. nr.44. 2003–2004. 'Erstatningsordning for krigsbarn og erstatningsordninger for romanifolk/tatere og eldre utdanningsskadelidende samer og kvener'.

Parliamentary Records/Meetings (Stortinget and Odelstinget)

O.tidene 25.5.1987, Sak nr 2, *Innstillingen fra sosialkomitéen om lov om kunstig befruktning* (Innst.O nr 60) Stortingsforhandlinger sesjon 1986–87, pp.308–45.
Stortingsforhandlinger 1988–89 nr 41.
Stortingstidende 10.6.1993, Sak nr 1, *Stortingsforhandlinger sesjon 1992–93*.
Stortinget. Meeting Monday 17 June 2002 (10:00), Sak nr 4, *Innstilling fra sosialkomitéen om evaluering av v om medisinsk bruk av bioteknologi* (Inst.S.nr 238 (2001–2002), jf. St.meld. nr 14 (2001–2002)).
Odelstinget. Meeting Tuesday 18 November kl 12.10 2003, Sak nr 1, *Innstilling fra sosialkomitéen m lov m humanmedisinsk bruk av bioteknologi m.m. (bioteknologiloven)* (Inst.O.nr 16 (2003–2004), jf OT.prp.nr 64 (2002–2003)). http://www–stortinget.no/otid/2003/o031118–01.html
Odelstinget. Meeting 24 May 2007 (13:08). http://www.stortinget.no/no/Saker–og–publikasjoner/Publikasjoner/Referater/Odelstinget/2006–2007/070524/1/. The discussion in Parliament was based on two documents: Innst. O. nr.62 (2006–2007) jf. Ot.prp. nr.26 (2006–2007).

Norwegian Lutheran Church Documents

1985. Kirkemøte 3/85: Menneskesyn og Menneskeverd.
1987. Kirkemøte 20/87: Kirkens engasjement for det ufødte liv.
1989. Kirkemøte 12/89: Bioteknologi og menneskeverdet.
1989. Kirkemøte 13/89: Abortlovgivningen.
Kirkerådet. 1989. *Mer Enn Gener. Utredning Om Bioteknologi Og Menneskeverd*. Oslo: Kirkerådet.
Bishops' Meeting 1989, 16/89, 'Etiske spørsmål vedrørende det ufødte liv'.

Published Works

Adrian, S. 2006. 'Nye skabelsesberetninger om æg, sæd og embryoner: Et etnografisk studie av skabelser på sædbanker og fertilitetsklinikker', Ph.D. thesis. Linkøping: University of Linkøping.
Appadurai, A. 2000[1996]. *Modernity At Large: Cultural Dimensions of Globalization*. Minneapolis: University of Minnesota Press.
Balen, F. van, and M. Inhorn. 2004. 'Interpreting Infertility: A View from the Social Sciences', in M. Inhorn and F. Van Balen (eds), *Infertility Around the Globe: New Thinking on Childlessness, Gender, and Reproductive Technologies*. Berkeley: University of California Press, pp.3–32.
Bangstad, S. 2009. *Sekularismens Ansikter*. Oslo: Universitetsforlaget.
Barnes, J. 1954. 'Class and Committees in a Norwegian Island Parish', *Human Relations* 7: 39–58.
Baune, Ø., O.J. Borge, S. Funderud, D. Føllesdal, G. Heiene, and L. Østnor. 2007. *Stamceller fra embryoer til forskning og terapi? Uttalelse fra en norsk forskningsgruppe*. Oslo: Unipub Forlag.
Becker, G. 2000. *The Elusive Embryo: How Women and Men Approach New Reproductive Technologies*. Berkeley: University of California Press.
Bernt, J.F. 2004. 'Vern av det befruktede egg: Hvor mye etisk politi?' *Morgenbladet*, 19–25 March. Retrieved 2 February 2008 from: http://www.Morgenbladet.no/apps/pbcs.dll/article?AID=/20040319/ARKIV/40319032.
Bestard, J. 1998. *Parentesco y modernidad*. Barcelona: Paidós.
Blyth, E., and J. Speirs. 2004. 'Meeting the Rights and Needs of Donor Conceived People: The Contribution of a Voluntary Contact Register', *Nordisk Sosialt Arbeid* 4(24): 318–30.
Bonnaccorso, M. 2009. *Conceiving Kinship: Assisted Conception, Procreation and Family in Southern Europe*. Oxford: Berghahn.
Borchgrevink, T. 2009. 'Instruire la nation: la religion dans la politique norvegienne', *Ethnologie Francaise* 39(2): 241–52.
Borneman, J. 2001. 'Caring and Being Cared For: Displacing Marriage, Kinship, Gender and Sexuality', in J. Faubion (ed.), *The Ethics of Kinship: Ethnographic Inquiries*, Lanham, MD: Rowman and Littlefield, pp.29–46.
Bradley, D. 1996. *Family Law and Political Culture: Scandinavian Laws in Comparative Perspective*. London: Sweet and Maxwell.
Brekke, O.A. 1995. *Differensiering og integrasjon: Debatten om bioteknologi og etikk i Norge*. Bergen: LOS Senteret.
Brekke, T. 2002. *Gud i norsk politikk: Religion og politisk makt*. Oslo: Pax Forlag.
Broberg, G. 2005. 'Scandinavia: An Introduction', in G. Broberg and N. Roll-Hansen (eds), *Eugenics and the Welfare State: Sterilization in Denmark, Sweden, Norway and Finland*. East Lansing: Michigan State University Press, pp.1–8.
Broberg, G., and N. Roll-Hansen (eds). 2005 [1996]. *Eugenics and the Welfare State: Sterilization in Denmark, Sweden, Norway and Finland*. East Lansing: Michigan State University Press.
Cadoret, A. 2009. 'The Contribution of Homoparental Families to the Current Debate on Kinship', in J. Edwards and C. Zalazar (eds), *European Kinship in the Age of Biotechnology*. Oxford: Berghahn, pp.79–96.

Campbell, B. 2007. 'Racialization, Genes and the Reinventions of Nation in Europe', in P. Wade (ed.), *Race, Ethnicity and Nation: Perspectives from Kinship and Genetics*. Oxford: Berghahn, pp.95–124.

Carsten, J. (ed.) 2000. *Cultures of Relatedness: New Approaches to the Study of Kinship*. Cambridge: Cambridge University Press.

Chemin, A. 2009. 'Don de gamètes: pour un régime restreint d'accès aux origines', *Le Monde*, 7 May 2009.

Christoffersen, S.A. 1986. 'Manipulasjon eller restriksjon? Etiske og teologiske synspunkter på bioteknologien', *Kirke og Kultur* 91: 100–19.

Cohen, L. 1999. 'Where It Hurts: Indian Material for an Ethics of Organ Transplantation', *Dædalus* 128(4): 135–65.

Collier, J., and S.J. Yanagisako (eds). 1987. *Gender and Kinship: Essays Toward a Unified Analysis*. Stanford, CA: Stanford University Press.

Dahl, H.F. 1986. 'Those Equal Folk', in S.R. Graubard (ed.), *Norden: The Passion for Equality*. Oslo: Norwegian University Press, pp.97–111.

Dalton, S. 2000. 'Non-biological Mothers and the Legal Boundaries of Motherhood: An Analysis of California Law', in H. Ragoné and F.W. Twine (eds), *Ideologies and Technologies of Motherhood: Race, Class, Sexuality, Nationalism*. New York: Routledge, pp.191–232.

Daniels, K. 1998. 'The Semen Providers', in K. Daniels and E. Haimes (eds), *Donor Insemination: International Social Science Perspectives*. Cambridge: Cambridge University Press, pp.76–104.

Daniels, K., and E. Haimes (eds). 1998. *Donor Insemination: International Social Science Perspectives*. Cambridge: Cambridge University Press.

Deech, R. 2003. 'Reproductive Tourism in Europe: Infertility and Human Rights', *Global Governance* 9: 425–32.

Deguchi, A. 2005. 'New Reproductive Technology and the Contemporary Family', in 'Toward the Construction of Death and Life Studies', *Bulletin of Death and Life Studies*, special issue, 1: 35–39.

Delaisi de Parseval, G. 2008. *Famille à tout prix*. Paris: Seuil.

Dolgin, J. 1992. 'Just a Gene: Judicial Assumptions about Parenthood', *UCLA Law Review* 40: 637–94.

——— 1997. *Defining the Family: Law, Technology, and Reproduction in an Uneasy Age*. New York: New York University Press.

——— 1999–2000. 'Choice, Tradition, and the New Genetics: The Fragmentation of the Ideology of Family', *Connecticut Law Review* 32: 523–66.

Donchin, A. 2010. 'Reproductive Tourism and the Quest for Global Justice', *Bioethics* 24(7): 323–32.

Døving, R. 2009. 'Le déjeuner norvégien: Le grand récit de la famille et de la nation', *Ethnologie Francaise* 39(2): 321–31.

Edwards, J. 2009. 'The Matter of Kinship', in J. Edwards and C. Salazar (eds), *European Kinship in the Age of Biotechnology*. Oxford: Berghahn, pp.1–18.

Edwards, J., S. Franklin, E. Hirsch, F. Price, and M. Strathern. 1999 [1993]. *Technologies of Procreation: Kinship in the Age of Assisted Conception*. London: Routledge.

Edwards, J., and C. Salazar (eds). 2009. *European Kinship in the Age of Biotechnology*. Oxford: Berghahn.

Edwards, R., and P. Steptoe. 1980. *A Matter of Life: The Story of a Medical Breakthrough*. London: Hutchinson.

Enger, C., and S. Hambro. 2010. 'Barn som bestilt', *Dagens Næringsliv Magasinet*, 27/28 February.
Finkler, K. 2001. 'The Kin of the Gene: The Medicalization of Family and Kinship in American Society', *Current Anthropology* 42(2): 235–63.
Fox, R. 1997 [1993]. *Reproduction and Succession: Studies in Anthropology, Law and Society*. New Brunswick, NJ: Transaction Publishers.
Franklin, S. 1995. 'Postmodern Procreation: A Cultural Account of Assisted Reproduction', in F. Ginsburg and R. Rapp (eds), *Conceiving the New World Order: The Global Politics of Reproduction*. Berkeley: University of California Press, pp.323–45.
——— 1997. *Embodied Progress: A Cultural Account of Assisted Conception*. London: Routledge.
——— 1999 [1993]. 'Making Representations: The Parliamentary Debate on the Human Fertilisation and Embryology Act', in J. Edwards, S. Franklin, E. Hirsch, F. Price, and M. Strathern, *Technologies of Procreation: Kinship in the Age of Assisted Conception*. London: Routledge, pp.127–65.
Franklin, S., and M. Lock. 2003. 'Animation and Cessation: The Remaking of Life and Death', in S. Franklin and M. Lock (eds), *Remaking Life and Death: Toward an Anthropology of the Biosciences*. Santa Fe, NM: School of American Research Press, pp.3–22.
Franklin, S., and S. McKinnon. 2001a. 'Relative Values: Reconfiguring Kinship Studies', in S. Franklin and S. Mckinnon (eds), *Relative Values: Reconfiguring Kinship Studies*. Durham, NC: Duke University Press, pp.1–25.
——— (eds). 2001b. *Relative Values: Reconfiguring Kinship Studies*. Durham, NC: Duke University Press.
Franklin, S., and H. Ragoné. 1998a. 'Introduction', in S. Franklin and H. Ragoné (eds), *Reproducing Reproduction: Kinship, Power, and Technological Innovation*. Philadelphia: University of Pennsylvania Press, pp.1–14.
——— (eds). 1998b. *Reproducing Reproduction: Kinship, Power, and Technological Innovation*. Philadelphia: University of Pennsylvania Press.
Frønes, I., A. Jensen, and A. Solberg. 1990. *Childhood as a Social Phenomenon: National Report – Norway*. Vienna: European Centre.
Fuller, C. 1994. 'Legal Anthropology, Legal Pluralism and Legal Thought', *Anthropology Today* 10(3): 9–12.
Geertz, C. 1983. 'Local Knowledge: Fact and Law in Comparative Perspective', in *Local Knowledge: Further Essays in Interpretive Anthropology*. New York: Basic Books, pp.167–234.
Ginsburg, F.D., and R.D. Rapp. 1995a. 'Conceiving the New World Order', in F. Ginsburg and R.D. Rapp (eds), *Conceiving the New World Order: The Global Politics of Reproduction*. Berkeley: University of California Press, pp.1–18.
——— (eds). 1995b. *Conceiving the New World Order: The Global Politics of Reproduction*. Berkeley: University of California Press.
Giæver, Ø. 2005. 'Eugenisk indikasjon for abort: en historisk oversikt', *Tidsskrift for Den norkse legeforening* 24(125): 3472–76.
Graubard, S. (ed.) 1986. *Norden: The Passion for Equality*. Oslo: Norwegian University Press.
Gullestad, M. 1985. *Livstil og likhet*. Oslo: Universitetsforlaget.
——— 1989. *Kultur og hverdagsliv*. Oslo: Universitetsforlaget.

—— 2001. 'Likhetens grenser', in M.E. Lien, H. Lidén, and H. Vike (eds), *Likhetens Paradosker: Antropologiske undersøkelser i det moderne Norge*. Oslo: Universitetsforlaget, pp.32–67.

—— 2002. 'Invisible Fences: Egalitarianism, Nationalism and Racism', *Journal of the Royal Anthropological Institute* 8(1): 45–64.

—— 2004. 'Normalising Racial Boundaries: The Norwegian Dispute about the Term Neger', *Social Anthropology* 13(1): 27–46.

Haimes, E. 1988. 'Secrecy: What Can Artificial Reproduction Learn from Adoption?', *International Journal of Law and Family* 2: 46–61.

—— 1990. 'Recreating the Family? Policy Considerations Relating to the "New" Reproductive Technologies', in M. McNeil, I. Varcoe, and S. Yearley (eds), *The New Reproductive Technologies*. Houndmills: Macmillan, pp.154–72.

—— 1992. 'Gamete Donation and the Social Management of Genetic Origins', in M. Stacey (ed.), *Changing Human Reproduction: Social Science Perspectives*. London: Sage, pp.119–47.

—— 1998. 'The Making of the "DI Child": Changing Representations of People Conceived through Donor Insemination', in K. Daniels and E. Haimes (eds), *Donor Insemination: International Social Science Perspectives*. Cambridge: Cambridge University Press, pp.53–75.

Halvorsen, M. 1998. 'The Act Relating to the Application of Biotechnology in Medicine with Particular Regard to Questions in Family Law', in A. Bainham (ed.), *The International Survey of Family Law*, pp.323–33.

Hargreaves, K. 2006. 'Constructing Families and Kinship through Donor Insemination', *Sociology of Health and Illness* 28(3): 261–83.

Harris, O. 1996. 'Inside and Outside the Law', in O. Harris (ed.), *Inside and Outside the Law: Anthropological Studies of Authority and Ambiguity*. London: Routledge, pp.1–18.

Hashiloni-Dovel, Y. 2007. *A Life (Un)Worthy of Living: Reproductive Genetics in Israel and Germany*. Dordrecht: Springer.

Hazekamp, J.T. 2005. 'When Things Become What You Call Them', *Human Fertility* 8(2): 79–80.

Hazekamp, J.T., and L. Hamberger. 1999. 'The Nordic Experience', in P. Brindsen (ed.), *Textbook of In Vitro Fertilization and Assisted Reproduction: The Bourn Hall Guide to Clinical and Laboratory Practice*. New York: Parthenon, pp.646–53.

Hellum, A. (ed.) 1993. *Birth Law*. Oslo: Scandinavian University Press.

Hellum, A., A. Syse, and H.S. Aasen (eds). 1990. *Menneske, naturen og fødselsteknologi: Verdivalg og rettslig regulering*. Oslo: Ad Notam.

Holy, L. 1996. *Anthropological Perspectives on Kinship*. London: Pluto Press.

Howell, S. 2001. 'Self-conscious Kinship: Some Contested Values in Norwegian Transnational Adoption', in S. Franklin and S. McKinnon (eds), *Relative Values: Reconfiguring Kinship Studies*. Durham, NC: Duke University Press, pp.203–23.

—— 2003. 'Kinning: The Creation of Life-trajectories in Adoptive Families', *Journal of the Royal Anthropological Institute* 9(3): 465–84.

—— 2006. *The Kinning of Foreigners: Transnational Adoption in a Global Perspective*. Oxford: Berghahn.

Howell, S., and M. Melhuus (eds). 2001. *Blod – tykkere enn vann? Betydninger av slektskap i Norge*. Bergen: Fagbokforlaget.

Humbyrd, C. 2009. 'Fair Trade International Surrogacy', *Developing World Bioethics* 9(3): 111–18.

Haave, P. 2000. *Sterilisering av tatere 1934–1977*. Oslo: Norges Forskningsråd.
Inhorn, M.C. 2002. 'The "Local" Confronts The "Global": Infertile Bodies and New Reproductive Technologies in Egypt', in M. Inhorn and F. van Balen (eds), *Infertility around the Globe: New Thinking on Childlessness, Gender, and Reproductive Technologies*. Berkeley: University of California Press, pp.263–82.
——— 2003. *Local Babies, Global Science: Gender, Religion, and In Vitro Fertilization in Egypt*. New York: Routledge.
——— 2007. 'Reproductive Disruptions and Assisted Reproductive Technologies in the Muslim World', in M. Inhorn (ed.), *Reproductive Disruptions: Gender, Technology, and Biopolitics in the New Millennium*. Oxford: Berghahn, pp.183–99.
Inhorn, M.C., and F. van Balen (eds). 2002. *Infertility around the Globe: New Thinking on Childlessness, Gender, and Reproductive Technologies*. Berkeley: University of California Press.
Inhorn, M., and D. Birenbaum-Carmeli. 2008. 'Assisted Reproductive Technology and Culture Change', *Annual Review of Anthropology* 37: 177–96.
Jenvin, O. 2008. 'Hvordan møter norsk byråkrati barna som kommer med cyberstorken?' in *Åpent møte om 'reproduksjonsturisme'*. Oslo: Bioteknologinemda and Nasjonalt Medisinsk Museum, pp.32–39.
Jordanova, L. 1995. 'Interrogating the Concept of Reproduction in the Eighteenth Century', in F.D. Ginsburg and R. Rapp (eds), *Conceiving the New World Order: The Global Politics of Reproduction*. Berkeley: University of California Press, pp.369–86.
Jørgensen, T. 2001. 'Til barnas beste: samværsrett og blodsbåndsforestillinger', in S. Howell and M. Melhuus (eds), *Blod – tykkere enn vann? Betydninger av slektskap i Norge*. Bergen. Fagbokforlaget, pp.119–42.
Kahn, S.M. 2002. 'Rabbis and Reproduction: The Uses of New Reproductive Technologies among Ultraorthodox Jews', in M. Inhorn and F. van Balen (eds), *Infertility around the Globe: New Thinking on Childlessness, Gender, and Reproductive Technologies*. Berkeley: University of California Press, pp.283–97.
Kaufman, S.R., and L.M. Morgan. 2005. 'The Anthropology of the Beginnings and Ends of Life', *Annual Review of Anthropology* 34: 317–41.
Kerridge, I.H., C.F.C. Jordens, R. Benson, R. Clifford, and R.A. Ankeny. 2010. 'Religious Perspectives on Embryo Donation and Research', *Clinical Ethics* 5: 35–45.
Kildal, N., and S. Kuhnle. 2005a. 'The Nordic Welfare Model and the Idea of Universalism', in N. Kildal and S. Kuhnle (eds), *Normative Foundations of the Welfare State: The Nordic Experience*. London: Routledge, pp.13–33.
——— (eds). 2005b. *Normative Foundations of the Welfare State: The Nordic Experience*. London: Routledge.
Konrad, M. 2005. *Nameless Relations: Anonymity, Melanesia and Reproductive Gift Exchange between British Ova Donors and Recipients*. Oxford: Berghahn.
Kooten Niekerk, N. van. 1994. *Teologi og bioetikk: Den protestantisk-teologiske vurdering af bioteknologien i Norden 1972–1991*. Aarhus: Aarhus Universitetsforlag.
Kopytoff, I. 1986. 'The Cultural Biography of Things: Commoditization as Process', in A. Appadurai (ed.), *The Social Life of Things: Commodities in Cultural Perspective*. Cambridge: Cambridge University Press, pp.64–91.
Kroløkke, C. 2012 forthcoming. 'From India with Love: Troublesome Citizens of Fertility Travel', *Cultural Politics*.

Kuhnle, S. 2001 [1994]. 'Velferdsstatens idégrunnlag i perspektiv', in A. Hatland, S. Kuhnle, and T.I. Romøren (eds), *Den norske velferdsstaten*. Oslo: Ad Notam Gyldendal, pp.9–31.
Kvande, L. 2008. 'Bilete av svangerskap–bilete av foster: Ultralyd diagnostikk i norsk svangerskaps omsorg 1970–1995'. Ph.D. thesis. Trondheim: Norwegian University of Science and Technology (NTNU).
Lappegård, T. 2000. 'New Fertility Trends in Norway' *Demographic Research* 2. Retrieved 15 March 2000 from: www.demographic-research.org/volumes/vol2/3.
——— 2007. 'Sosiologiske forklaringer på fruktbarhetsendring i Norge i nyere tid', *Sosiologisk Tidsskrift* 15: 55–71.
Lavik, N.J. 1998. *Rasismens intellektuelle røtter*. Oslo: Tano Aschehoug.
Lefaucher, N. 2004. 'The French "Tradition" of Anonymous Birth: The Lines of the Argument', *International Journal of Law, Policy and the Family* 18(3): 319–42.
Leira, A. 1992. *Welfare States and Working Mothers: The Scandinavian Experience*. Cambridge: Cambridge University Press.
——— 1996. *Parents, Children, and the State: Family Obligations in Norway*. Oslo: Institute for Social Research.
Leirvik, O. 2004. 'The Current Debate about Religious Education and Freedom of Religion in Norway'. Retrieved 15 March 2011 from: http://folk.uio.no/leirvik/oslocoalition/leirvik0902htm.
Lien, M.E., H. Lidén, and H. Vike. 2001. *Likhetens Paradokser: Antropologiske undersøkelser i det moderne Norge*. Oslo: Universitetsforlaget.
Lien, M.E., and M. Melhuus. 2007. 'Introduction', in M.E. Lien and M. Melhuus (eds), *Holding Worlds Together: Ethnographies of Knowing and Belonging*. Oxford: Berghahn, pp.ix–xxiii.
——— 2009. 'La Norvège: Vues de l'interieur', *Ethnologie Francaise* 39(2): 197–207.
Lock, M. 2002. *Twice Dead: Organ Transplants and the Reinvention of Death*. Berkeley: University of California.
Lundin, S. 1997. *Guldägget: Föräldreskap i biomedicinens tid*. Lund: Historiska Media.
Løvset, J. 1951. 'Artificial Insemination: The Attitude of Patients in Norway', *Fertility and Sterility* 2(5): 414–29.
McNeill, M., I. Varcoe, and S. Yearley (eds). 1990. *The New Reproductive Technologies*. New York: St Martin's Press.
Marcus, G. 1998. *Ethnography through Thick and Thin*. Princeton, NJ: Princeton University Press.
Martin, E. 1991. 'The Egg and the Sperm: How Science Has Constructed a Romance Based on Stereotypical Male-Female Roles', *Signs* 16(3): 485–501.
Melby, K., A. Pylkkänen, B. Rosenbeck, and C.C. Wetterberg. 2000. *The Nordic Model of Marriage and the Welfare State*. Copenhagen: Nordic Council of Ministers.
——— 2006. *Äktenskap och politik i norden ca 1850–1930*. Göteborg: Makadam Förlag.
Melhuus, M. 1997. 'The Troubles of Virtue: Values of Violence and Suffering in a Mexican Context', in S. Howell (ed.), *The Ethnography of Moralities*. London: Routledge, pp.178–202.
——— 2003. 'Exchange Matters: Issues of Law and the Flow of Human Substances', in T.H. Eriksen (ed.), *Globalisation: Studies in Anthropology*. London: Pluto Press, pp.170–97.

——— 2005. '"Better Safe than Sorry": Legislating Assisted Conception in Norway', in C. Krohn-Hansen and K. Nustad (eds), *State Formation: Anthropological Perspectives*. London: Pluto Press, pp.212–33.

——— 2007. 'Procreative Imaginations: When Experts Disagree on the Meanings of Kinship', in M.E. Lien and M. Melhuus (eds), *Holding Worlds Together: Ethnographies of Knowing and Belonging*. Oxford: Berghahn, pp.37–56.

Melhuus, M., and S. Howell. 2009. 'Adoption and Assisted Conception: One Universe of Unnatural Procreation', in J. Edwards and C. Salazar (eds), *European Cultures of Kinship in the Age of Biotechnology*. Oxford: Berghahn, pp.144–61.

Merry, S.E. 1992. 'Anthropology, Law and Transnational Processes', *Annual Review of Anthropology* 21: 357–79.

Mo, E., M. Seliussen, L.M. Irgens and K. Gåsemyr. 2006. *Register over nemdbehandlede svangerskapsavbrudd*. Nasjonalt Folkehelseinstituttt og Medisinsk Fødselsregister.

Molne, K. 1976. 'Donorinseminasjon: en oversikt og et materiale', *Tidsskrift for Den Norske Lægeforening* 17/18: 982–86.

Moore, S.F. 1987. 'Explaining the Present: Theoretical Dilemmas in Processual Ethnography', *American Ethnologist* 14(4): 727–36.

Morgan, L. 1992. 'When Does Life Begin? A Cross-cultural Perspective on the Personhood of Fetuses and Young Children', in W.A. Haviland and R.J. Gordon (eds), *Talking about People: Readings in Contemporary Cultural Anthropology*. Mountain View: Mayfield Publishing, pp.28–38.

——— 1998. 'Ambiguities Lost: Fashioning the Fetus into a Child in Ecuador and the United States', in N. Scheper-Hughes and C. Sargent (eds), *Small Wars: The Cultural Politics of Childhood*. Berkeley: University of California Press, pp.58–74.

——— 2002. '"Properly Disposed of": A History of Embryo Disposal and the Changing Claims on Fetal Remains', *Medical Anthropology* 21: 247–74.

Mulkay, M. 1997. *The Embryo Research Debate: Science and the Politics of Reproduction*. Cambridge: Cambridge University Press.

NAVF. 1986. *Tverrfaglig konferanse om ufrivillig barnløshet: nye løsninger, nye problemer*. Oslo: NAVF, Sekretariat for kvinneforskning.

Nielsen, T.H., A. Monsen, and T. Tennøe. 2000. *Livets tre og kodenes kode; Fra genetikk til bioteknologi. Norge 1900–2000*. Oslo: Gyldendal Akademisk.

Plesner, I.T. 2005. 'Relgionspolitiske modeller og dilemmaer', in D. Søderlind (ed.), *Farvel til Statskirken? En debattbok om kirke og stat*. Oslo: Humanist Forlag, pp.13–36.

Porqueres i Gené, E. 2009. *Défis contemporains de la parenté*. Paris: Editions de l'Ecole des Hautes Etudes.

Porqueres i Gené, E., and J. Wilguax. 2009. 'Incest, Embodiment, Genes and Kinship', in J. Edwards and C. Salazar (eds), *European Kinship in the Age of Biotechnology*. Oxford: Berghahn, pp.112–27.

Pottage, A. 2004. 'The Fabrication of Person and Things', in A. Pottage and M. Mundy (eds), *Law, Anthropology, and the Constitution of the Social: Making Persons and Things*. Cambridge: Cambridge University Press, pp.1–39.

——— 2007. 'The Socio-legal Implications of the New Biotechnologies', *Annual Review of Law and Social Science* 3: 321–44.

Rabinow, P. 1999. *French DNA: Trouble in Purgatory*. Chicago: University of Chicago Press.

Ragoné, H. 1994. *Surrogate Motherhood: Conceptions in the Heart*. Boulder, CO: Westview Press.

Riksaasen, G. 2001. 'To mammaer, går det an? En annerledes familieplanlegging', in S. Howell and M. Melhuus (eds), *Blod – tykkere enn vann? Betydninger av slektskap i Norge*. Bergen: Fagbokforlaget, pp.99–118.

Rivière, P. 1985. 'Unscrambling Parenthood: The Warnock Report', *Anthropology Today* 1(4): 2–7.

RMF. 1983. *Etiske retningslinjer ved kunstig befruktning (AID) og in vitro fertilisering (IVF)*. Oslo: NAVF/Rådet for Medisinsk Forskning, Utvalg for forskningsetikk.

Roberts, E. 2007. 'Extra Embryos: The Ethics of Cryopreservation in Ecuador and Elsewhere', *American Ethnologist* 34(1): 181–99.

Roll-Hansen, N. 1999. 'Eugenics in Scandinavia after 1945: Change of Values and Growth in Knowledge', *Scandinavian Journal of History* 24(2): 199–213.

——— 2005. 'Norwegian Eugenics: Sterilization as Social Reform', in G. Broberg and N. Roll-Hansen (eds), *Eugenics and the Welfare State: Sterilization Policy in Denmark, Sweden, Norway and Finland*. East Lansing: Michigan State University Press, pp.151–94.

Rommetveit, K. 2005. 'Bioteknologiloven i en helserettslig sammenheng', *Kritisk Juss* 31(2): 168–83.

Rose, N. 2007. *The Politics of Life Itself. Biomedicine, Power, and Subjectivity in the Twenty-first Century*. Princeton, NJ: Princeton University Press.

Rønne-Pettersen, E. 1951. *Provrörsmänniskan: En studie i moderne magi*. Stockholm: Bokförlaget Biopsykologi.

Salazar, C. 2009. 'Are Genes Good to Think With?', in J. Edwards and C. Salazar (eds), *European Kinship in the Age of Biotechnology*. Oxford: Berghahn, pp.179–96.

Sandberg, K.A. 2009. 'Barns rettslig stilling i likekjønnende parforhold', in H. Aune, O.K. Fauchald, K. Lilleholt and D. Michaelsen (eds), *Arbeid og rett*. Oslo: Cappellen Akademiske, pp.547–65.

Sandemose, A. 1952. 'Unnfanget i løgn', *Årstidende* 2: 24–86; 3: 39–53.

Schneider, D. 1980 [1968]. *American Kinship: A Cultural Account*, rev. edn. Chicago: University of Chicago Press.

——— 1984. *A Critique of the Study of Kinship*. Ann Arbor: University of Michigan Press.

Seip, A.-L. 1994. *Veiene til velferdsstaten*. Oslo: Gyldendal Norsk Forlag.

Shimazono, S. 2005. 'Foreword', in 'Towards the Construction of Death and Life Studies', *Bulletin of Death and Life Studies*, special issue, 1: 7–10.

Shore, C. 1992. 'Virgin Births and Sterile Debates: Anthropology and the New Reproductive Technologies', *Current Anthropology* 33(3): 295–314.

Simpson, B. 2001. 'Making "Bad" Deaths "Good": The Kinship Consequences of Posthumous Conception', *Journal of the Anthropological Institute* 7(1): 1–17.

——— 2004. 'Impossible Gifts: Bodies, Buddhism and Bioethics in Contemporary Sri Lanka', *Journal of the Royal Anthropological Institute* 10(4): 839–59.

Sirnes, T. 1997. *Risiko og meining: Mentale brot og meningsdimensjonar i industri og politikk*. Bergen: Universitetet i Bergen, Institutt for Administrasjon og Organisasjonsvitenskap.

Skodje, H.T. 2011. '9 av 10 aborterer Downs', *Aftenposten*, 18 March. Retrieved 1 February 2012 from: https://web.retriever-info.com/services/archive.html.

Solbakk, J.H. 2005. 'Etiske utfordringer i norsk biolovgivning', *Kritisk Juss* 31(2): 184–99.

Solberg, B. 2003a. 'Etikken i å si nei til sorteringssmafunnet', *Genialt* 2: 20–23.

—— 2003b. 'Den nye bioteknologiloven–ikke til barnets beste likevel?' *Nytt Norsk Tidsskrift* 20(3): 316–24.
Solerød, M. 2008. 'Endelig fedre', *Tikdsskrift for jordmødre*, 6. Retrieved 1 February 2012 from: http://www.jordmorforeningen.no/tj/Tidsskrift-for-jordmoedre/Tema/2008/Dnj-jubilerer/Endelig-fedre.
Spallone, P. 1989. *Beyond Conception: The New Politics of Reproduction*. Cambridge, MA: Bergin and Garvey.
Spilker, K.H. 2008. 'Assistert slektskap: Biopolitikk i reproduksjonsteknologiens tidsalder', Ph.D. thesis. Trondheim: Norwegian University of Science and Technology (NTNU).
Spilker, K.H., and M. Lie. 2007. 'Gender and Bioethics Intertwined: Egg Donation within the Context of Equal Opportunity', *European Journal of Women's Studies* 14(4): 327–40.
Stanworth, M. (ed.) 1987. *Reproductive Technologies: Gender, Motherhood and Medicine*. Cambridge: Polity Press.
Starr, J., and J.F. Collier. 1989. 'Dialogues in Legal Anthropology', in J. Starr and J.F. Collier (eds), *History and Power in the Study of Law: New Directions in Legal Anthropology*. Ithaca, NY: Cornell University Press, pp.1–28.
Stenius, H. 1997. 'The Good Life is a Life of Conformity: The Impact of Lutheran Tradition on Nordic Political Culture', in Ø. Sørensen and B. Stråth (eds), *The Cultural Construction of Norden*. Oslo: Scandinavian University Press, pp.161–71.
Stolcke, V. 1986. 'New Reproductive Technologies, Same Old Fatherhood', *Critique of Anthropology* 6: 5–32.
—— 2009. 'A proposito del sexo', *Política y Sociedad* 46(1/2): 43–55.
Storrow, R.F. 2005. 'Quests for Conception: Fertility Tourists, Globalization and Feminist Legal Theory', *Hastings Law Journal* 57(2): 295–329.
Strathern, M. 1981. *Kinship at the Core: An Anthropology of Elmdon, a Village in North-west Essex in the Nineteen-sixties*. Cambridge: Cambridge University Press.
—— 1992a. 'The Meaning of Assisted Kinship' in M. Stacey (ed.), *Changing Human Reproduction: Social Science Perspectives*. London: Sage, pp.148–69.
—— 1992b. *After Nature: English Kinship in the Late Twentieth Century*. Cambridge: Cambridge University Press.
—— 1995. 'Displacing Knowledge: Technology and the Consequences for Kinship', in F.D. Ginsburg and R. Rapp (eds), *Conceiving the New World Order: The Global Politics of Reproduction*. Berkeley: University of California Press, pp.346–68.
—— 1999. *Property, Substance, and Effect: Anthropological Essays on Persons and Things*. London: Athlone Press.
—— 2005. *Kinship, Law and The Unexpected: Relatives Are Always a Surprise*. Cambridge: Cambridge University Press.
Stråth, B. 2005. 'The Normative Foundations of the Scandinavian Welfare States in Historical Perspective', in N. Kildal and S. Kuhnle (eds), *Normative Foundations of the Welfare State: The Nordic Experience*. London: Routledge, pp.34–51.
Sundby, J. 1989. 'Subfertiltiet og barnløshet: Et utvalg av norske kvinner', *Tidsskrift for Den Norske Lægeforening* 19–21: 1996–98.
—— n.d. 'Forplantningsteknologi og infertilitet: et skjæringspunkt mellom individuelle behov og samfunsmæssig styring', *Nytt Forum for Kivnneforskning* 2.
Sundby, H., and G. Guttormsen. 1989. *Infertilitet*. Otta: TANO

Syse, A. 1990. 'Fra egg til mennesket: medisinske og rettslige problemstillinger knyttet til fosterets rettstuvikling', in A. Hellum, A. Syse, and H.S. Aasen (eds), *Menneske, natur og fødselsteknologi: Verdivalg og rettslig regulering*. Oslo: Ad Notam, pp.89–106.
—— 1993. *Abortloven: Juss og verdier*. Oslo: Ad Notam Gyldendal.
Sætre, S. 2010. 'Sæd story', *Morgenbladet*, 8–14 January, pp.8–11.
Sørensen, Ø., and B. Stråth. 1997a. 'The Cultural Construction of Norden', in Ø. Sørensen and B. Stråth (eds), *The Cultural Construction of Norden*. Oslo: Scandinavian University Press, pp.1–24.
—— (eds). 1997b. *The Cultural Construction of Norden*. Oslo: Scandinavian University Press.
Thompson, C. 2005. *Making Parents: The Ontological Choreography of Reproductive Technologies*. Cambridge, MA: MIT Press.
Thorkildsen, D. 1997. 'Religious Identity and Nordic Identity', in Ø. Sørensen and B. Stråth (eds), *The Cultural Construction of Norden*. Oslo: Scandinavian University Press, pp.138–60.
Tjørnhøj-Thomsen, T. 1998. 'Tilblivelseshistorier: Barnløshed, slægtskab og forplantningsteknologi i Danmark', Ph.D. thesis. Copenhagen: Department of Anthropology, University of Copenhagen.
Tranøy, K.E. 1989. 'Den nye fruktbarhetsteknologien: nye og gamle rettigheter', *Tidsskrift for Rettsvitenskap* 2: 112–26.
Trägårdth, L. 1997. 'Statist Individualism: On the Culturality of the Nordic', in Ø. Sørensen and B.Stråth (eds), *The Cultural Construction of Norden*. Oslo: Scandinavian University Press, pp.252–85.
Vike, H., H. Lidén, and M.E. Lien. 2001. 'Likhetens virkeligheter', in M.E. Lien, H. Lidén, and H. Vike (eds), *Likhetens Paradokser: Antropologiske undersøkelser i det moderne Norge*. Oslo: Universitetsforlaget, pp.11–31.
Wade, P. 2007. *Race, Ethnicity and Nation: Perspectives from Kinship and Genetics*. Oxford: Berghahn.
Warnock Committee. 1985. *A Question of Life: The Warnock Report on Human Fertilisation and Embryology*. Oxford: Blackwell.
Witoszek, N. 1997. 'Fugitives from Utopia: The Scandinavian Enlightenment Reconsidered', in Ø. Sørensen and B. Stråth (eds), *The Cultural Construction of Norden*. Oslo: Scandinavian University Press, pp.72–90.
Yanagisako, S., and C. Delaney (eds). 1995. *Naturalizing Power: Essays in Feminist Cultural Analysis*. New York: Routledge.
Østnor, L. (ed.) 2008. *Stem Cells, Human Embryos and Ethics: Interdisciplinary Perspectives*. Springer.
Aasen, H.S. 1990. 'Noen perspektiver knyttet til dagens utvikling innen prenatal diagnostikk', in A. Hellum, A. Syse and H.S. Aasen (eds). *Menneske, natur og fødselsteknologi: Verdivalg og rettslig regulering*. Oslo: Ad Notam, pp.27–42.

INDEX

A
adoption
  child of their own, significance of 29, 30, 32–3, 34, 36, 37–8, 40, 41, 42, 43, 44
  legislative process, law and 52, 53, 54, 55, 56, 66, 113–14
  motherhood and legislation 84–5
  preferences for 28
  transnational adoption 28, 43–4
Adrian, S. 136
Al Aqeel, Dr Aida Ibrahim M. 149n37
Alvheim, John 96
anonymity
  anonymous sperm donation 107
  motherhood and 75–6, 77, 78
Appadurai, S. 9
artificial, connotations of 111
artificial conception (*kunstig befruktning*) 15, 66, 111, 141n33
artificial insemination by donor (AID) 3, 15, 50, 51, 52, 53, 54, 55, 56, 57, 64, 66, 69, 76, 101, 111
  arguments for and against 51–5
  biological arguments about 52, 53, 54
  psychological arguments about 52–3
Artificial Procreation Act (1987) 2–3, 50–51, 57–8, 60–61, 63, 76, 135n4, 139n1
  motherhood 76
assisted conception (*assistert befruktning*) 2–3, 4, 10, 11, 14, 15, 17
  magic of 53–4
  motherhood and 73–4
  studies of 7–8
assisted reproductive technologies (ARTs) 3, 5, 7, 111, 115, 116
  involuntary childlessness (and the involuntary childless) and 110, 121

Association for Fertility and Childlessness
(*Ønskebarn*) 138n1, 145n28
Associaton for the Involuntary Childless (FUB) 23, 24, 25, 27, 29, 40, 122, 124, 126, 138n1, 139n9, 145n28

B
Baby M 7
Bangstad, S. 138n39
Barnes, J. 137n33
Becker, G. 90, 146n9
Bernt, J.F. 97, 108
Bestard, J. 136n11
bilateral kinship system 44
bioethics, religion and 102–3
biogenetic connectedness
  child of their own, significance of 26, 27, 28, 44
  legislative process, law and 113
  motherhood and 73, 85
biogenetics, definitive quality of 113
biological connectedness 118
biology and rights, notions of 81
biomedicine, choice and 90
biopolitics 1, 6, 9, 13, 15, 21
biotechnology
  fears concerning uses of 97
  potentialities of 9–10
Biotechnology Act (1994) 3–4, 12, 13, 63, 65, 84, 91, 93, 135n3, 135n5, 136n8, 137n23, 140n9, 143n53
  opposition to revisions to, mood of 68–9
  proposed amendment (2011) 98
  revision of (2003) 65–7
  revision of (2007) 67–9
Biotechnology Act (2003), parliamentary debate on revision 93–8
Biotechnology Act (2007) 107, 111, 117, 120

Biotechnology Advisory Board 62, 82, 123, 124
birth mother, concealment of identity of 119
Blood, Diane 109
Blyth, E. and Spiers, J. 140n15
bodies (and parts), commercialization of 7
Bohlin, Kjell 77
Bonaccorso, M. 136n14
Bondevik, Kjell Magne 65
bonding, concerns about 26–7
Borchgrevink, T. 19
Borneman, J. 84
Bradley, D. 140n10
Brekke, O.A. 135n3, 138n39, 142n40, 142n42
Broberg, G. 102
Broberg, G. and Roll-Hansen, N. 99, 147n21
Brown, Louise 4
Bruce, Donald 149n37
Brundtland, Gro Harlem 60

C

Cadoret, A. 81
Campbell, B. 146n6
Carlsen, Pastor Ingvald B. 100
Carsten, J. 136n13
Catholic Church 18, 100, 149n39
Centre of Equality 77, 78
Centre Party 62, 65, 67, 95
Chemin, A. 145n19
child of their own, significance of 11–12, 23–45, 73, 113–14
    adoption 29, 30, 32–3, 34, 36, 37–8, 40, 41, 42, 43, 44
        preferences for 28
    bilateral kinship system 44
    biogenetic connectedness 26, 27, 28, 44
    bonding, concerns about 26–7
    conjugal relation 32, 43
        personal experience of importance of 38–40, 41–3
    egg donation 32, 36, 37, 38, 39, 40, 42, 43, 44
    emotional turmoil, personal experience of 38–40
    emotions of involuntary childlessness 23–4, 25–6
    flaunting success in childbearing 24
    future baby, imagination of 44
    imagined sameness 28, 43, 45
    infertility
        treatments for, cycles of 29–30
    infertility, sensitiveness of 31
    kinship
        assumptions of 26
        self-conscious discourse on 27
    male infertility 31–2
    mourning for childlessness 23–4
    mutual problem of childlessness 31–2
    natural creation 26
    'own child,' meanings of notion of 25–6, 29–30
    personal experiences 32–43
    pregnancy
        personal experience of desire for 34–6, 36–8
        significance of 29, 30, 35, 37, 39, 40, 43
    process of assisted conception, difficulties of 30–32
    procreation and filiation, ties between 45
    professional couple, personal experience of 32–4
    relatedness 26, 27, 28, 32, 40, 43, 44, 45
    reproductive technologies and options provided 24–5, 28, 35, 42, 43
    research parameters 24–5
    sameness, issue of 27–9
    sharing involuntary childlessness with others 31–2
    sperm donation 32, 35, 37, 38, 40, 42, 43, 44
    surrogacy 29, 34, 38, 44
    time, intercession of 25–6
    transnational adoption 28, 43–4
    see also involuntary childlessness
Children's Act (1981, amended 1998) 73, 77, 81, 117
choice
    denial of, problem of 94–5, 95–6
    rights of 96–7
Christian Democratic/Conservative coalition 86
Christian Democratic Party 61, 64, 65, 67, 68, 93, 96, 104, 106
Christoffersen, S.A. 149n44
Church Council (1989) 102–3
codification of law 9, 57–63
cohabitation 45, 57, 72, 131
Cohen, L. 146n7
Collier, J. and Yanagisako, S.J. 136n12
conception
    meaning (and visibility) of 48–9
    procreation and issues of 49
conjugal relation
    child of their own, significance of 32, 43
    motherhood and 73
    personal experience of importance of 38–40, 41–3
Conservative Party 61, 64, 65, 67, 68, 95
CRYOS (sperm bank in Denmark) 123

D

Dahl, H.F. 16, 17

Dalton, S. 145n32
Daniels, K. 144n5, 144n6
Daniels, K. and Haimes, E. 140n15
Deech, R. 3
Delaisi de Parseval, G. 107, 123, 136n6, 136n14
denial of choice, problem of 94–5, 95–6
deoxyribonucleic acid (DNA) 77, 83, 85, 86, 87, 89, 107, 123
discrimination 42, 44, 63, 75, 77, 92, 94, 105–6
disruptive effects 80–82
divorce 5, 29, 39, 52, 72, 131, 141n18
Dolgin, J. 7, 21, 80, 109, 136n17, 145n26
Donchin, A. 110, 150n1
Duesund, Woie 96
Døving, R. 138n43

E
Edwards, J. et al. 136n11, 136n13
Edwards, Jeanette 1, 6, 71, 136n10
Edwards, R. and Steptoe, P. 4
egg donation
 child of their own, significance of 32, 36, 37, 38, 39, 40, 42, 43, 44
 motherhood 73, 74–8, 79, 80, 82, 84, 87
embryo
 human embryo, status of 7–8
 moral status 92, 106–7
 social field 91
emotions
 involuntary childlessness (and the involuntary childless) 23–4, 25–6
 turmoil of, personal experience of 38–40
equality, choice and 113, 114
Equality, voice of Centre for 77–8
ethics
 dilemmas and publicity 104–6
 ethical policing 97–8
 ethical publicity 92
Ethics Committee 62
eugenics 2, 15, 16, 91–2, 98–102

F
family
 core values of 56
 notion of 86
 significance for childless 30, 86
fatherhood 15
 father is established through marriage (*pater est quem nuptiae demonstrant*) 72–3
 see also paternity
filiation 2, 7, 15, 16
Finkler, K. 107, 136n13
Fox, R. 7, 80, 136n11

Franklin, S. and Lock, M. 108, 137n22
Franklin, S. and McKinnon, S. 136n12, 136n13
Franklin, S. and Ragoné, H. 21, 136n13
Franklin, Sarah 7, 47, 74, 78, 91, 108, 111, 139n2, 149n36
Frønes, I., Jensen, A. and Solberg, A. 151n3
FUB *see* Association for the Involuntary Childless (FUB)
Fuller, C. 136n16
future baby, imagination of 44

G
Geertz, Clifford 8, 10, 12, 136n16
Gene Technology Act (1993) 135n3, 142n40
Germany 3, 89, 98–9, 136n15
Ginsburg, F.D. and Rapp, R.D. 1, 5, 49, 136n13
Giæver, Ø. 101, 147n14, 147n21, 148n23, 148n27
globalization of reproductive technologies 109–10
Gullestad, M. 17, 28, 137n32, 138n40

H
Haimes, E. 11, 136n11, 145n18
Halvorsen, M. 141n35
Hargreaves, K. 140n15
Harris, O. 11, 109
Hashiloni-Dovel, Y. 136n15
Hazekamp, J.T. 144n68
Hazekamp, J.T. and Hamberger, L. 60
Hellum, A. 74
Hellum, A., Syse, A. and Aasen, H.S. 74
Helsinki Declaration (1964) 101
Holy, L. 136n11
Howell, S. 10, 11, 28, 43, 44, 81, 139n11
Howell, S. and Melhuus, M. 135n1
human dignity 95–6
human embryo, status of 7–8
human rights 76–7
Humbyrd, C. 110
Høie, Bent 95
Høybråten, Dagfinn 65, 93, 94, 95
Hålogaland, Bishop of 76
Haave, Per 101, 147n22

I
identity, biologization of 15–16
illegitimacy 49, 53, 72, 137n19, 140n12
imaginations
 imagined sameness 28, 43, 45
 legislative process and 8–10
in vitro fertilization (IVF) 3, 4, 10, 11, 15, 24, 25, 29, 37, 39, 42, 44, 48, 51, 56, 61, 62, 63, 64, 66, 68
 first Norwegian baby born 57

infertility
  male infertility 31–2
  perceptions of 128–9
  sensitiveness of 31
  treatments for, cycles of 29–30
Inhorn, Marcia C. 4, 5, 7, 47, 136n7
Inhorn, M.C. and Birenbaum-Carmeli, D. 136n13
Inhorn, M.C. and van Balen, F. 49, 136n13
intracytoplasmic sperm injection (ICSI) 37, 40, 41, 42, 136, 136n7, 138, 138n3
inviolability of motherhood 71–87
involuntary childlessness (and the involuntary childless) 13–14, 48, 49, 55, 81, 86, 124–5, 126–8, 139–40n5, 142n49
  assisted reproductive technologies (ARTs) and 110, 121
  community of 10–12, 151n2
  definition of 136n18
  emotions of 23–4, 25–6
  family, significance for 30, 86
  infertility in, perceptions of 128–9
  technological innovation and, concerns about 59–60
  see also child of their own, significance of
Italy 3, 132, 136n14

J
Jenvin, Odd 82, 83, 84, 85, 86, 87, 107, 116, 117, 145n28
Jewish societies 7–8, 103
Jordanova, Ludmilla 1
Jørgensen, T. 72

K
Kahn, S.M. 7, 145n32
Kaufman, S.R. and Morgan, L.M. 136n15
Kildal, N. and Kuhnle, S. 137n34, 137n35, 137n36
kinship 4–6, 7
  assumptions of 26
  bilateral kinship system 44
  biological connectedness 7
  blood relations and 72
  categories of 80
  knowledge, information and 107–8
  law and
    family, core values of 56
    maintenance of certainty, objective of 48
    uncomfortable relationship between 48
  legislative process, law and 113
  public negotiations of motherhood 71–87
  self-conscious discourse on 27
Kinship, Law and the Unexpected (Strathern, M.) 47

Kirkemøtet 101, 148n33, 148n35
Kirkerådet 102, 148n35
knowledge
  accessibility of 3
  ethics, policy and 97–8
  information, kinship and 107–8
Knudsen, Grethe 79, 80, 82
Kopytoff, I. 8
Kristoffersen, A. 96, 146n10
Kroløkke, C. 146n33
Kuhnle, S. 137n36
Kvande, L. 93, 146n8, 147n13, 147n18, 149n41

L
Labour Party 62, 65, 67, 96
Lappegård, T. 28, 72, 131, 132, 133
Lavik, N.J. 100, 147n19, 148n25
Lefaucher, N. 145n25
legislative process, law and 2, 14–16, 47–69
  Acts and revisions 50–51
  adoption 52, 53, 54, 55, 56, 66, 113–14
  artificial insemination by donor (AID)
    arguments for and against 51–5
    biological arguments about 52, 53, 54
    psychological arguments about 52–3
  Artificial Procreation Act (1987) 2–3, 50–51, 57–8, 60–61, 63, 76, 135n4, 139n1
  assisted conception, magic of 53–4
  biogenetics, definitive quality of 113
  biogenetic relatedness 113
  Biotechnology Act (1994) 3–4, 12, 13, 63, 65, 84, 91, 93, 135n3, 135n5, 136n8, 137n23, 140n9, 143n53
    proposed amendment (2011) 98
    revision of (2003) 65–7
    revision of (2007) 67–9
  Biotechnology Act (2007) 107, 111, 117, 120
  Church and law 102–3
  codification of law 9, 57–63
  conception
    meaning (and visibility) of 48–9
    procreation and issues of 49
  historical perspective 55–6
  imaginations and 8–10
  Kinship, Law and the Unexpected (Strathern, M.) 47
  kinship and law 113
    family, core values of 56
    maintenance of certainty, objective of 48
    uncomfortable relationship between 48
  legal sensibility 10
  'Local Knowledge: Fact and Law in Comparative Perspective' (Geertz, C.) 8–9, 10

marriage, centrality of institution of 53, 55
maternity 113
opposition to revisions to Biotechnology Act, mood of 68–9
parliamentary debates 66–7, 68
parliamentary report (1953) 51, 56
paternity 113
reproductive technologies 47, 48–9, 51, 58, 69
science and technology, perspectives on developments in 48
secrecy, concerns about 54–5
sexual relations 49, 52
uncertainties within 116–20
unitary motherhood, idea of 113
Leira, A. 26, 53, 141n18
Leirvik, O. 19
Liberal Party 65
Lien, M.E. and Melhuus, M. 16, 20, 105
Lock, M. 1, 150n2
Lundin, S. 136n14
Lutheran Church (*Kirkemøtet*, Church meeting) 101, 148n33, 148n35
(*Kirkerådet*, Church Council) 102, 148n35
Løvset, J. 51, 54, 141n22

M

male infertility 31–2
male same-sex parenthood 82–7
Marcus, G. 6
marriage, centrality of institution of 53, 55
Marriage Act (1918) 147n21
Marriage Act (1991) 72
Marriage Act (2008) 72, 86, 117, 143n69, 145n31
Martin, E. 144n4
maternity 118
   legislative process, law and 113
   see also motherhood
Melby, K. et al. 140n10, 141n19
Melhuus, M. 3, 5, 9, 50, 58, 75, 139n3, 143n57
Melhuus, M. and Howell, S. 11, 137n19, 137n27, 140n15, 143n57, 143n66, 145n26
Merry, S.E. 8, 136n16, 136n17
Ministry of Health and Social Affairs 62, 63, 64, 79
Mo, E. et al. 147n17
Molne, K. 141n22, 141n30, 141n32
Molne, K. and Khan, J. 57
Moore, S.F. 122
moral ambiguity, dispute and 3
Morgan, L. 136n15
motherhood 15, 118
   adoption legislation 84–5

anonymity 75–6, 77, 78
Artificial Procreation Act (1987) 76
assisted conception 73–4
biogenetic connectedness 73, 85
biology and rights, notions of 81
Biotechnology Advisory Board 82
birth mother, concealment of identity of 119
Children's Act (1981, amended 1998) 73, 77, 81, 117
cohabitation 72
conjugal relation 73
discrimination 77
disruptive effects 80–82
divorce 72
egg donation 73, 79, 80, 82, 84, 87
eggs and sperm 74–8
Equality, voice of Centre for 77–8
family, notion of 86
father is established through marriage (*pater est quem nuptiae demonstrant*) 72–3
human rights 76–7
illegitimacy 72
inviolability of 71–87
kinship
   blood relations and 72
   categories of 80
   public negotiations of motherhood and 71–87
male same-sex parenthood 82–7
mater semper certa est 72–3, 79–80
mother is always certain 72–3, 79–80
National Medical Museum 82
National Population Register 83
nature and nurture 77, 84
'own child,' meanings of notion of 73
parenthood, establishment of 71–4
pater vero, male same-sex parenthood 82–7
paternity
   male same-sex parenthood 82–7
   named sperm and 78
Property, Substance and Effect (Strathern, M.) 71
reproductive technologies 73–4, 78
sperm donation 74, 75, 78, 79, 80, 87
state bureaucracy, reality and 83–4
surrogacy 82–3, 84–5
Theology, attitude of Norwegian School towards artificial insemination 76
unitary motherhood, idea of 113
mourning for childlessness 23–4
Mulkay, M. 104, 147n15, 149n36
Muslim societies 7–8, 103
Møller, Katti Anker 148n23
Møre, Bishop of 76

N

National Bureau of Statistics 131
National Medical Museum 82
National Population Register 83
natural creation 26
nature and nurture 4, 27, 42, 44, 52, 140n14
    motherhood and 77, 84
Nazi Germany 98–9, 102, 104
new reproductive technologies (NRTs) 3
Niekerk, Kooten 149n36
Nielsen, T.H., Monsen, A. and Tennøe, T. 98, 99, 100, 101, 147n14, 147n19, 148n24
Norway
    Biotechnology Advisory Board 62, 82, 123, 124
    equality, notion in 16–21
    notion of 'sorting society' (*sorteringssamfunnet*) in 92–3, 94, 105–6
    reproductive technologies, legislative wariness on 2
    royal bloodline in 14
    technological innovation, uneasiness about 9
Norwegian Christian Doctors Association 76
Norwegian Lutheran Church 84, 94, 100, 101, 102, 148n30
Norwegian Research Council 135n2
Norwegian School of Theology 76
Nuremberg Code (1947) 101

O

Olsen, Gunn 96
Ombudsperson 77
Organization for Economic Cooperation and Development (OECD) 133
Orheim, Karita Bekkemellem 96
'own child,' meanings of notion of
    child of their own, significance of 25–6, 29–30
    motherhood and 73

P

parenthood, establishment of 71–4
parliamentary debates 66–7, 68
parliamentary report (1953) 51, 56
pater est 72–3
pater vero, male same-sex parenthood 82–7
paternalism 96
paternity 119
    establishment of 119–20
    legislative process, law and 113
    male same-sex parenthood 82–7
    named sperm and 78
Peek, Laura 137n26
personal experiences 32–43
Plesner, I.T. 138n59

policy
    knowledge, ethics and 97–8
    and science, relationship between 100–101
The Politics of Life Itself (Rose, N.) 91
Porqueres i Gené, E. 136n13
Porqueres i Gené, E. and Wilgaux, J. 82
Pottage, A. 9, 10, 12, 15, 137n21
precautionary principle 2, 9, 12–13, 113
pregnancy
    personal experience of desire for 34–6, 36–8
    significance of 29, 30, 35, 37, 39, 40, 43
preimplantation genetic diagnosis (PGD) 3, 15–16, 64, 66, 67, 68, 69, 114
    'sorting society' and 89, 92, 93, 102, 103, 104, 107, 108
prenatal diagnosis (PND) 3, 15–16, 64, 66, 67, 69, 97, 114, 148n34
    'sorting society' and 89, 92, 93, 95, 102, 103, 104, 107, 108
procreation 2, 5, 11, 15
    filiation and, ties between 45
procreative choice 87, 120
procreative universe 4, 9, 110, 115, 117
Progress Party 64, 65, 67, 96
Property, Substance and Effect (Strathern, M.) 71
public healthcare 42, 61, 62, 90, 98, 142n46
public–private divide 2, 4

R

Rabinow, P. 9, 89, 136n14
racial hygiene 98–9
Ragoné, H. 136n13, 145n32
relatedness 26, 27, 28, 32, 40, 43, 44, 45
religion
    bioethics and 102–3
    role of 7–8
reproduction
    reproductive technologies 2
    social transformation and 1–2
reproductive choice 107, 114–15
    'sorting society' and 89–92
reproductive technologies
    confrontation of 95
    globalization of 109–10
    impact of 6–7
    legislative process, law and 47, 48–9, 51, 58, 69
    legitimacy (and limits) of 111
    local-global dialectic of 109–12
    motherhood 73–4, 78
    and options provided for childless 24–5, 28, 35, 42, 43
    reception in Norway 110–11
    'sorting society' (*sorteringssamfunnet*) 90–91
    studies of 7–8

research parameters 24–5
Rights of the Child, UN Convention on the (1991) 66, 81, 117–18
Riksaasen, G. 146n35
Rivière, P. 7
Roaldkvam, Sidsel 138n2
Roberts, E. 7
Roll-Hansen, N. 99, 100, 101, 149n43
Romany people 101, 148n30, 148n32
Rose, N. 91, 92, 99, 106, 146n4, 146n5
Rønne-Pettersen, E. 54, 141n21, 141n23

S

Salazar, C. 48
sameness, issue of 27–9
Sandberg, K.A. 145n30
Sandemose, A. 54
Scandinavian countries 3, 92, 99, 102, 137n30, 140n10, 141n18
Schneider, D. 136n12
science
  policy and, relationship between 100–101
  technology and, perspectives on developments in 48
Second World War 100, 101
secrecy, concerns about 54–5
secular societies 7–8
Seip, A.-L. 147n19, 147n20
selective abortion 16
  'sorting society' and 95, 104–5
sexual relations 49, 52
Shimazono, S. 7
Shore, C. 7
Simpson, B. 7, 92, 109
Sirnes, T. 64, 142n38, 142n42, 142n47, 142n49, 149n38, 149n45
Skodje, H.T. 147n17
Smedal, Olaf 135n1
Smidt, Bishop 53
Smidt, Johannes 140n8
Social Affairs, Committee of 60–61, 124
Socialist Left Party 62, 65, 67
society
  biotechnologies and 115
  moral fabric of 112–13
  technology and 95–6
socio-cultural potency of biological information 108
Solbakk, J.H. 149n37, 149n44
Solberg, B. 93, 97, 98
Solerød, M. 145n28
'sorting society' (sorteringssamfunnet) 2, 4, 15, 16, 89–108, 114
  anonymous sperm donation 107
  bioethics, religion and 102–3
  biomedicine, choice and 90
  biotechnology, fears concerning uses of 97
  Catholic Church 100
  choice, rights of 96–7
  Church Council (1989) 102–3
  denial of choice, problem of 94–5, 95–6
  embryo
    moral status 92, 106–7
    social field 91
  ethical policing 97–8
  ethical publicity 92
  ethics, dilemmas and publicity 104–6
  eugenics 91–2, 98–102
  human dignity 95–6
  kinship, knowledge and information 107–8
  knowledge
    ethics and policy 97–8
    information and kinship 107–8
  law and Church 102–3
  Lutheran Church (Kirkemøtet, Kirkerådet) 94, 100, 101, 102, 148n33, 148n35
  Norwegian notion of 92–3, 94, 105–6
  Nuremberg Code (1947) 101
  parliamentary debate on revised Biotechnology Act (2003) 93–8
  paternalism 96
  policy
    knowledge and ethics 97–8
    and science, relationship between 100–101
  The Politics of Life Itself (Rose, N.) 91
  preimplantation genetic diagnosis (PGD) 89, 92, 93, 102, 103, 104, 107, 108
  prenatal diagnosis (PND) 89, 92, 93, 95, 102, 103, 104, 107, 108
  racial hygiene 98–9
  religion and bioethics 102–3
  reproductive choice 89–92
  reproductive technologies 90–91
    confrontation of 95
  science and policy, relationship between 100–101
  selective abortion 95, 104–5
  society, technology and 95–6
  socio-cultural potency of biological information 108
  sterilization laws 99–101
Spain 74, 132, 146n6
Spallone, P. 136n11
sperm donation 14, 15
  motherhood and 74, 75, 78, 79, 80, 87
  prohibition of 74
  significance for childless 32, 35, 37, 38, 40, 42, 43, 44
Spilker, K.H. 86

Spilker, K.H. and Lie, M. 74
Stabel, Byråsjef Carl 140n8
Stanworth, M. 136n11
Starr, J. and Collier, J.F. 136n16
state bureaucracy, reality and 83–4
Stenius, H. 18, 20
sterilization laws 99–101
Stolcke, V. 136n11, 144n4
Strathern, M. 1, 15, 21, 26, 47, 49, 71, 104, 107, 136n11, 136n12, 136n13, 137n28, 138n44
Stråth, B. 138n42
Sundby, H. and Guttormsen, G. 136n18
Sundby, J. 136n18
Sundfør, Hans 140n8
surrogacy 2, 7, 8, 11, 15, 18, 118–19
   birth mother, concealment of identity of 119
   motherhood and 82–3, 84–5
   significance for childless 29, 34, 38, 44
Syse, A. 101
systematic selection 105, 114
Sørensen, Ø. and Stråth, B. 17, 18, 21, 137n30, 137n35

T

technological innovation, concerns about 59–60
   see also reproductive technologies
Theology, attitude of Norwegian School towards artificial insemination 76
Thompson, C. 2, 5, 109, 136n13
Thorkildsen, D. 19, 20
time, intercession in childlessness of 25–6
Tjørnhøj-Thomsen, T. 136n14
Trägårdh, L. 18, 20

transnational adoption 28, 43–4
Troms Chief Administrative Officer 77

U

Ulfnes, Kari Anne 151n1
UN Convention on the Rights of the Child (1991) 66, 81, 117–18
uncertainties within legislative process 116–20
unitary motherhood, idea of 113
United States 6, 7

V

Vike, H., Lidén, H. and Lien, M.E. 16, 17, 20, 21, 137n31
Volden, Kari Ann 84

W

Wade, P. 146n6
Warnock Committee (1985) 7, 140n7
Witoszek, N. 19
World Wars (I and II) 98
   see also Second World War

Y

Yanagisako, S. and Delaney, C. 136n12

Z

Zoloth, Laurie 149n37

Ø

Østnor, L. 103, 149n36

Aa

Aasen, H.S. 74, 148, 149n42, 153